REBOOTING Childhood

REBOOTING Childhood

HOW TO BALANCE SCREEN TIME, BUILD RELATIONSHIPS, AND HELP KIDS THRIVE

KIRSTY NOLAN, MS

REBOOTING Childhood

HOW TO BALANCE SCREEN TIME, BUILD RELATIONSHIPS, AND HELP KIDS THRIVE

BY KIRSTY NOLAN, MS

The names and details of the people and situations described in this book have been changed or presented in composite form in order to ensure the privacy of those with whom the author and organization have served.

Published by

www.chaddock.org

Cover design and interior layout by pearcreative.ca
Edited by coastalediting.com

ISBN: 979-8-9918469-4-3 (Print)

To any adult out there caring for a child or young person, who is also trying to navigate the digital storm, finding balance is HARD. I see you and I salute you.

To the children and young people who have taught me so much, especially my own daughter, Kenzie, and my stepdaughter, Carlie; I am so thankful. I understand that my presence is more important than any digital luxury this world develops. I promise to keep doing better for you.

To the people who have supported me on this literary journey: my husband, my family, and my wonderful and exceptionally knowledgeable colleagues, I appreciate you.

To Dad and Dave: I wouldn't be where I am today without your love and guidance. Wish you were both here to read this.

And finally, a special mention to the late Dr. Debbie Reed, who always encouraged me to reach for those big, hairy, audacious goals—this one is for you.

CONTENTS

Chaddock
Every Child Deserves a Chance

ACKNOWLEDGMENTS

This book, and The Developmental Trauma and Attachment Program (DTAP®) that our concepts are rooted in, represents years of learning, reflection, and growth at Chaddock. It is the result of collective dedication to providing hope and healing to children and families who need both.

We want to recognize that *all* our work is shaped by the interactions we have with everyone that crosses our path—children, families, staff, donors, volunteers, community partners, and supporters—we thank you for your energy, your insight, and for sharing your stories with us.

I also want to personally acknowledge the professional contributions made by President and CEO of Chaddock, Matt Obert, Dr. Caelan Soma and Kaylee Morford, without whom this book may not have gotten off the ground. And, to the incredibly talented Lizzie Moon, whose delightful illustrations have brought this book to life. Finally, to my ever patient and gracious supervisor, Molly Bainter, for believing in my ability to make this happen.

AUTHOR'S NOTE

When you look back on the days of your childhood, what do you remember most? If you're anything like me, your free time was spent outside, riding bicycles, playing, and adventuring through the vast wilderness that was your neighborhood. Weekends consisted of walks with the dog, outings, board games, and chores—always connected with others. Being bored was a part of growing up, and if you wanted something new or extra, you had to work hard, and wait, to get it.

I consider myself one of the lucky ones. The days of my childhood were spent exploring the world around me with the ones I loved the most. Was it perfect? No, but there was food on the table and a roof over my head, and while we were comfortable financially, we were by no means wealthy. Instead, what made my life truly rich was the relationships and connections I had with family and friends. I cannot recall a time in my childhood when I ever felt unloved, unworthy, unsafe, isolated, or alone. Even during the times that my parents had to work, I was taken care of by neighbors, and family friends, or I was with them at work, engaged in an activity, such as drawing, reading, writing, or simply just *being*. Life revolved around real-world friendships and conflicts often confined by geographical distance and how far you were willing to cycle.

Through these relationships and real-life experiences, I learned the social, emotional, and physical skills needed to be a mostly (because who's perfect?) healthy and happy child, adolescent, and adult.

My memories of primary school consist of playtime, toys, fun, colors, shapes, messes, and music. I don't think I ever saw a television set in school until I was about 9 years old (and yes, I'll admit, we all got a little excited when we saw the teacher roll it in on its cart). In the later years of high school, we were lucky if we got to watch a video, or (hooray!) a DVD to substantiate whatever topic we were learning at the time.

At home we had one small television and a radio that was always playing in the kitchen. If we wanted to speak to friends, we had to call them on the landline phone—hoping that they would pick up and we wouldn't have to talk to their parents, or even worse, their older sibling! If you were lucky, you had a camera but had to patiently await the pictures that were coming in the mail until one of your friends got a Polaroid.

Technology as it is today did not exist, and I am so thankful for that.

That being said, I would be a hypocrite if I said that I am not also thankful for the digital era we now live in. I cannot deny the impact that cutting-edge technology, the internet, social media, and cell phones have had on my life. As a person who has emigrated from her home country, the value of these advances is immeasurable: thousands of miles of ocean may separate us but at the press of a button and a matter of seconds, I can see my mother; I can type a few letters into a box and connect with friends that I might not have seen for years; and I can see pictures of my nephews and their latest achievements as the moments happen.

But as a mother to a toddler, a stepmother to a teen, and a professional who supports children, families, and caregivers, I can, and do, see the consequences of this high-tech age. While we have the capability to be more connected than at any other point in history, it feels like we are more disconnected than we have ever been. Not only that but we have young children who cannot regulate their brains and bodies because they have been "co-regulated" by the bright lights and fast images of a screen, older kids who collapse without instant gratification, and teens who are trying to navigate both the real world and the virtual one without a break from either.

Parents and other caring adults find themselves battling their children as they seek to regain control (or at the very least find balance), all while trying to avoid epic meltdowns or defiant behavior; often it feels easier to just give in (which, by the way, I've done that, a lot).

So, if you're feeling overwhelmed by screens, confused by all the "technical" terms, or just exhausted from trying to figure it out, take a breath. You are not alone. You're exactly who this book is for.

With patience, practice, and perseverance (and a little support from us), you *can* reclaim your connection with your children and balance the positives of our digital world.

And don't ever forget, the most important thing in front of your child's eyes is *you*.

Let's walk this road together,
Kirsty & The Chaddock Team

ABOUT THIS BOOK

There is no doubt that the way we live and connect with each other has changed, as has the way we raise our children. And let's be honest, parenting today can sometimes feel like trying to build a sandcastle during a tsunami.

What used to be squabbles over toys are now arguments about screen time. Instead of "Mom, can I go outside?" it's "Can I finish one more level?" And we probably shouldn't even get into how many dinner conversations have been hijacked by the addictive pull of a glowing rectangle.

The endless swirl of screens, notifications, videos, influencers, reels, stories, how-tos, guidelines, comments, likes, and tweets has reshaped childhood and parenting in ways many of us never could imagine and can leave even the most well-intentioned caregiver feeling overwhelmed, confused, and exhausted.

But here's the thing; you are not failing. You are raising children in a digital age that none of us were trained for, and most of you are doing the best you can with what you have. And you're doing it all with heart.

WHAT'S REALLY GOING ON?

Well, as you might have noticed, our children have moved from a "play-based childhood" to a "phone-based childhood" and no longer spend their free time playing with friends in the real world, engaging in what Jonathan Haidt, author of *The Anxious Generation*, describes as *"embodied, synchronous and one-to-one or one-to-several"* interactions (Haidt, 2024, p.9). Instead, the average tween spends up to six hours a day staring at screens. Teens? More like nine. And for many parents, those numbers aren't just statistics; they're real-life battles fought before breakfast.

In fact, many childhood interactions now no longer require another human body to be present, and with the onset of AI, sometimes not even another human brain!

The trouble is, in order to become physically, emotionally, and socially healthy, kids still need what they've always needed—love, connection, emotional safety—and all of these things require another real-life human being.

To quote a wonderful colleague, Steve Zwolak,

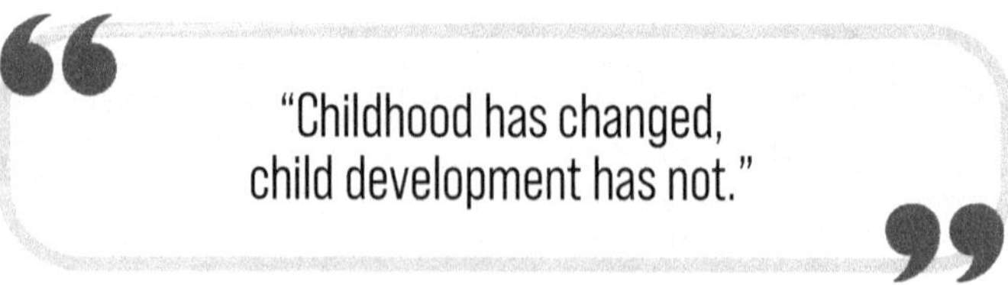

This is what we will refer to throughout this book as *digital disruption*: the process by which digital media, screens, technology, the internet, smartphones, etc. can interrupt healthy physical, social, and emotional development.

SO, WHAT DO WE DO NOW?

Well, the answer isn't to "put down your phone forever." And it's definitely not about shaming you for handing over a tablet so you could take a shower or drink your coffee while it's still hot. Because, whether we want to admit it or not, advanced technology such as screens, smartphones, and the internet are a part of our world and will continue to be.

But it *is* about balance, boundaries, and most importantly, relationships.

This book was created for real families, real classrooms, and real professionals who are trying—every single day—to connect with kids in a world that seems determined to distract them.

We have blended what we know from Chaddock's Developmental Trauma and Attachment Program (DTAP®) with what we feel from journeying with children and families for more than 170 years. This is where clinical insight meets kitchen-table wisdom.

It's grounded in the science of how kids grow, how their brains develop, and how relationships heal. But it's written in a way that feels more like a conversation than a college lecture—because we've sat across from parents with tear-filled eyes and teachers who've reached their limits. We know what hopelessness and exhaustion look like; we've lived them too.

This book is your road map for:

- understanding the science behind how screens affect your child,
- learning how to respond when digital disruption strikes,
- using our DTAP® concepts to manage behavior *and* maintain connection, and
- swapping excess for balance.

It's part guide, part companion, part cheerleader. And it's all grace.

HOW TO USE THIS BOOK

It took a long time for this book to come together. We wanted to make sure we gave you the knowledge you need to implement the tools successfully, while also providing you with easy, go-to strategies that you can use right away.

There's no "right way" to read this book. We designed it to be skimmable, flippable, and full of quick take-aways, but in order to make the most of it, here are our suggestions:

1. Start Where You Are

If you've got toddlers and tech tantrums, bedtime battles with tweens or TikTok-obsessed adolescents and need immediate answers, jump to part 2. We get it; sometimes you just need a solution.

However, if you have the time (and mental energy!) and want to learn about *why* we recommend certain strategies, begin with the chapters in part 1 and use them as a resource to support you when you start using the tools in part 2. We've found that having this knowledge can make it easier to follow through and persevere when initially it might feel like things still aren't working.

A QUICK NOTE!

Throughout the book you will find specific words in bold type: **Attunement, Window of Tolerance, Structure and Nurture, The Importance of Play, Verbal Responses, Sitting in the Yuck, Problem-Solving and Abstract Thinking, Independent Regulation** and **Natural and Logical Consequences**. These are the concepts that underpin our DTAP® treatment approach and understanding them will be the key to your success—you can find all of them defined in detail in chapter 5.

And remember, you can always come back to them whenever you need a refresher.

2. Look for the Highlights

Throughout the book we have sprinkled in helpful tidbits that are easy to find:

Brain Bites: Simple science that makes sense (and sticks) and shows what's going on behind the scenes.

Real Talk: Relatable quotes from parents, caregivers and professionals who have been (and are living) in the trenches.

Quick Tip(s): Try-it-tonight strategies and scripts (including what to say and how to respond when things don't go as planned).

Next-Level Knowledge: Links that will guide you to earlier chapters if you want or need to know more about the concept we are talking about.

3. Keep It Real

We get it, life is messy. Sometimes the strategies won't go as planned. Sometimes you'll read a section and think, *Wow, I wish I knew this last week*. It's okay. This book isn't about being perfect; it's about being purposeful. And that's more than enough.

4. Print It, Post It, Pass It On

Many parents have told us, "I literally ripped that page out and stuck it on the fridge." Good! That's exactly the kind of book this is. If something resonates, share it. Text it to a friend. Quote it in a parent group. Our goal is to create ripples of connection that go beyond the pages of this book.

At the end of each chapter, and often after important sections of information, you'll find some fridge-worthy facts that you might want to keep handy.

And remember, you are not alone in this.

Wherever you are on this journey, however you use this book, and whether you read one chapter or all of them, we hope it feels like a trusted friend walking alongside you.

SOME IMPORTANT CONSIDERATIONS

Wouldn't life be easier if all our kids were the same? Had the same experiences? Came from the same backgrounds? Well, maybe. But it would also be incredibly boring, rather than rich and diverse (and who doesn't love a little challenge now and then?).

It's important to remember that aside from digital disruption there are many other factors that can impact a child or young person's social, emotional, and behavioral functioning.

For example:

Family Make-Up/Dynamics

Firstly, we recognize that caregiving takes many forms—by parents, grandparents, relatives, foster or adoptive parents, and even teachers. For simplicity, the term "parent" has been used throughout this book to include *all* who care for children.

With this in mind, it's especially important for all those caring for a child to be on the same page and consistent in their approach when it comes to managing any challenges that may arise from digital disruption. You're unlikely to see positive change and progress if the adults aren't speaking the same language.

Developmental Trauma and Relational (Attachment) Disruptions

A child or adolescent who has a history of developmental trauma or disrupted attachment relationships (i.e., significant inconsistencies, trauma, or neglect in primary caregiving) may experience challenges relating to their social, emotional, and behavioral functioning. We must also recognize that brain functioning might be compromised, meaning that areas of the brain needed for complicated tasks such as decision-making, **problem-solving**, impulse-control, logic, and reasoning may not be fully accessible—many children and young people with these histories will be working from their Downstairs Brain (something you'll learn more about soon). Luckily, because this is our area of expertise, all of the strategies you will learn in the latter part of this book were specifically chosen because they allow us to meet the child where they are at in their brain.

It is possible that digital disruption will exacerbate these challenges. However, it is also possible (as with all children and young people) that there may be times when technology can be used positively.

Neurodiversity

Children who experience neurodiversity, much like those who do not, can gain both positive and negative effects from technology and digital media. Some of the potential challenges associated with this population include exposure to bullying or misunderstandings due to difficulties with social cues and communication; worsening of pre-existing symptoms such as shorter attention spans and difficulty concentrating on non-preferred activities; and vulnerability to sensory overload from bright screens, loud sounds, or fast-paced visuals due to heightened sensory systems.

We recommend that caregivers of children who are neuro-diverse ensure that the strategies suggested in part 2 are individually tailored to each child's specific set of needs, in a way that works for them and their family.

Cultural and Societal Norms

We understand that all families, children, and youth are different, and nobody is perfect. There are many other factors that can, and will, impact how we manage technology, screens, social media, and digital experiences in our children's lives, including our culture, values, our own upbringing and experiences of being parented, our inner circle of support, and the outer pressures of society.

We recognize that we cannot comment on all these diversities within this book, and so we encourage you to take our suggestions and do what works best for you, in your setting, with your children and young people. Be reassured: Even small changes can make a big difference!

Part I

UNDERSTANDING WHAT'S REALLY GOING ON

Chapter 1

THE BRAIN UNDER CONSTRUCTION

You Leveled Up! Congrats, You Have Reached Level 2!

Have you ever wondered how your child can go from learning to tie their shoes one minute to throwing said shoes across the room at you the next? Or why your teenager can solve complex algebra equations but not remember to do the dishes like you asked them to just two minutes ago? Welcome to the world of brain development.

Kids' and adolescents' brains are under construction until late into their twenties, and just like a building, some parts are move-in ready, while others require a lot more framing before you can even begin drywalling. Some parts even need renovating after they have been built, especially if something disrupts the build process. And even though development typically happens in a chronological order, it doesn't mean that our teenagers won't behave like toddlers sometimes (which is completely normal, in case you were wondering). Higher level cognitive skills such as empathy, **abstract thinking** and impulse control are the last to show up to the housewarming party. It's not bad behavior; it's just biology.

Brain Bite: The brain develops from bottom to top, inside to outside.

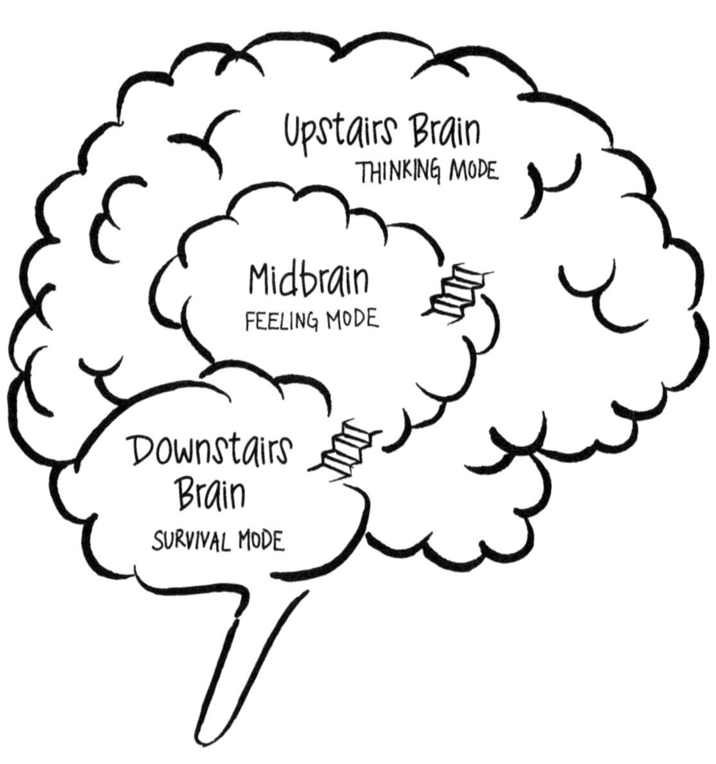

The Downstairs Brain = Survival Mode.

The oldest part of the brain, which functions as our survival mechanism through fight, flight, or freeze responses. It's the body's built-in alarm system, designed to alert us quickly, without asking questions first. This is the part that says, "Empty stomach! We might die from starvation! Must eat!"

The Midbrain = Feeling Mode.

The part of the brain that links our downstairs sensations to our upstairs logic—you can think of this as the landing, with stairs going up and stairs going down. It helps us begin to put words to our physical and emotional responses so that we can communicate them to others. This is the part that says, "I'm hangry! Someone feed me before I bite!"

The Upstairs Brain = Thinking Mode.

Home to logic, reasoning, empathy, organization, **problem-solving**, decision-making and many other critical skills. This is the part that says, "Going to the grocery store on an empty stomach wasn't such a good idea."

Brain Bite: It is important to understand that if a child or young person (or anyone of any age for that matter) is in their Downstairs Brain, they will be unable to access their Upstairs Brain (and all its important functions) until they feel safe and connected to people around them.

Survival Mode Says:

"Empty stomach! We might die from starvation! Must eat!"

Feeling Mode Says:

"I'm hangry! Someone feed me before I bite!"

Thinking Mode Says:

"Ugh! Going to the store on an empty stomach wasn't a very good idea!"

Real Talk

Take Ava. She's 5. Here's Ava's mom describing a recent interaction they had:

"I asked Ava to switch off her tablet, explained to her that we had to leave and go to a doctor's appointment and we were already running late. She had a huge meltdown, was screaming and threw the tablet on the floor. I told her again that we were running late, but she just carried on screaming."

Without meaning to, Ava's mom had overwhelmed Ava's regulation system by asking her to switch off the tablet (this is a BIG CHANGE for a 5-year-old brain), causing Ava to jump into her Downstairs Brain and giving her no access to her Upstairs Brain, making any kind of logical thinking impossible, thus the huge meltdown.

Without access to her Upstairs Brain (not to mention the fact that Ava is only 5 and hasn't yet mastered the skill of understanding time), no amount of reasoning from her mom was going to help Ava feel emotionally safe and regulated.

So, what did Ava really need from her mom during that moment?

Recognition that her Downstairs Brain was in charge.
"Phew I know this is hard right now, I'm here."

Reassurance that big feelings don't mean bad behavior.
"It's okay to feel upset; I'll take the tablet so it doesn't get broken."

Regulation with a calm, connected presence.
"Let's take a breath and figure this out together."

Relational Repair (when everyone is calm): *"I know it's hard when you have to stop doing something fun. It's not okay to throw things though, as they might hurt somebody. Next time I will help you by giving you lots of reminders, how about that?"*

Quick Tip(s): Before you correct, connect. A calm, "I'm here" when a child is melting down often does more than a five-minute lecture. Remember, less talking, more connection.

I get it, this brain science stuff can be a little overwhelming, and you're probably wondering, "How does this relate to my child's meltdown when she gets to play her tablet all day?" Or "How does this apply to my kid who slams the door and sulks when I make them turn their phone in at night?" Perhaps even, "How does this help me talk to my teenager about accessing explicit content online?"

Well, if we know what part of the brain our child or adolescent is functioning in, we can stop reacting impulsively and tailor our response in a way that:

a. they can understand;
b. doesn't make the situation worse; and
c. maintains the relationship, while also maintaining expectations.

For example, if we can recognize (using **attunement**) that our 5-year-old's meltdown is because they are in their Downstairs Brain (and outside of their **window of tolerance**), we will know not to use logic but instead co-regulate and meet basic needs using a calm voice, physical proximity, and short, clear messages (this is known as balancing **structure and nurture**). The reasoning part comes later.

Similarly, if we understand that our kid is most likely functioning in their Midbrain when it's time to hand in their phone at night, we can **sit in the yuck** with them and **validate** their emotions with empathy *"I get it, your friends are your whole world right now and I'm asking you to let go of that for a bit; that must be really hard,"* while also maintaining the limits using **play**, *"Don't forget that you'll see them tomorrow, and if you don't get a good night's sleep you turn into a bit of an angry bear; that's why the phone gets put up."*

And finally, if we remember that our teenager has access to their (mostly) developed Upstairs Brain when we have a difficult conversation, for example, about watching explicit content online, we can help them use **abstract thinking**, **problem-solving**, and decision-making skills successfully to manage their behavior (developing their ability to **independently regulate**) without shaming them *and* still maintain expectations through the use of **natural and logical consequences**.

Brain Bite: The brain and body learn through meltdowns and small, infrequent stressors, allowing for the formation of new neural pathways (that's brain connections to you and me) and a healthy stress response system, but only in the context of a safe, healthy relationship with a caregiver.

Throughout this construction process, you are the project manager and your ability to bring calm will ensure its success. However, there's no requirement to memorize every blueprint in your head. Instead, your role is to show up to help guide construction while maintaining safety, especially when the scaffolding starts to sway.

So, when your toddler screams after hearing, "No," your kid lashes out when you tell them it's time to pause the video, or your teen shuts down after being embarrassed on social media, remember: They aren't giving you a hard time; they are *having* a hard time.

And guess what? The most powerful response you can provide is your continuous presence through it all.

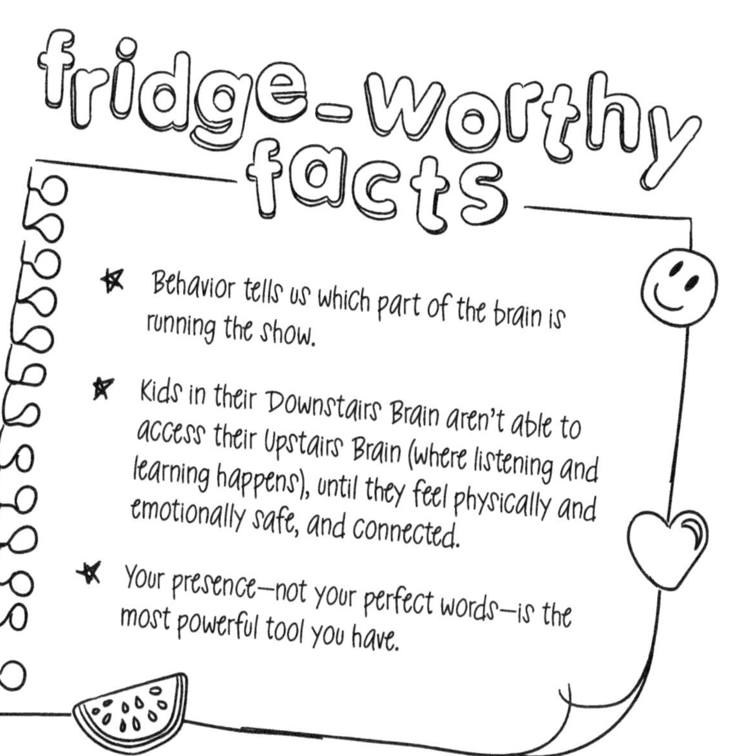
fridge-worthy facts
Behavior tells us which part of the brain is running the show.
Kids in their Downstairs Brain aren't able to access their Upstairs Brain (where listening and learning happens), until they feel physically and emotionally safe, and connected.
Your presence—not your perfect words—is the most powerful tool you have.

Chapter 2

THE JOURNEY OF DEVELOPMENT

Your Sim Has Advanced to the Next Stage of Life: New Traits Unlocked!

UNDERSTANDING YOUR CHILD'S DEVELOPMENT

Your child does not arrive with an instruction manual (oh how we can wish!), but they do come equipped with a developmental road map, and this chapter functions as your guide to understanding the journey they are on.

As they grow through different ages and stages, children will experience fresh challenges and opportunities, as well as distinct quirks. The uniqueness of each child requires understanding about what is driving behavior to reduce frustration and allow us to respond effectively—this is known as **attunement**.

Our exploration will focus on essential aspects of development that span from birth until the late teenage years. These developmental milestones go beyond physical abilities to establish the basic framework for emotional control and higher-level social functioning as well as personal identity and resilience.

Our journey starts with infancy and toddlerhood, when brain development happens at an extraordinary speed.

DEVELOPMENT IN THE EARLY YEARS (BIRTH–AGE 5)

"Where Little Minds Take Giant Leaps"

Brain Bite: The National Scientific Council on the Developing Child (2007) tells us that the developmental changes that take place during the early years (0–5) are the *greatest* and most *rapid* compared to any other period of development.

So, if you ever wondered what's really going on in your toddler's head when they decide to drink out of the dog bowl, now you know.

And it's the reason we will spend a little more time focused on this age group than the others when we talk about development.

Children aged 0–5 go through some of the biggest physical milestones of their lives. (Remember when you couldn't wait for them to start walking only to find that you had to baby-proof the entire house?) They also achieve some of the biggest (and most vital) social and emotional milestones, which, much like their physical counterparts, require connection, **attunement**, reciprocation, confirmation, and correction in order for a child to successfully master them. *Very few, if any*, of these milestones can be mastered independently.

These milestones are represented in the graphic below. You will see that within each age bracket one key milestone has been identified (those highlighted in bold): Attachment, Psychological Home Base, Object Constancy, Mastery, and Power vs. Powerlessness. The other milestones within the age range all play a part in the healthy achievement of the key milestone.

0-1	1-2	2-3	3-4	4-5
Reciprocity, Nurture and Attunement	Locomotion	Autonomy	"Why" Questions	Peer Group
Symbiosis	**Psychological Homebase**	"No" and "Mine"	Cause and Effect Relationship	Cooperative Play and Dramatic Play
Regulation of State	Transitional Object	Toileting	Symbolic Play, Sequences and Elaborate Play	Identity and Gender
Attachment	Language Bridge	Body Integrity	Same and Different	**Power vs. Powerlessness**
Differentiation	Early Level Symbolic Play	Parallel Play	**Mastery**	Body Integrity
Object Permanence	Separation	Integration of Positive and Negative Effects	Monster Dreams	Birth, Death and Injury
Pointing, Labeling	Ambivalence	**Object Constancy**		Fantasy vs. Reality

Note. Adapted from: *Creating Schools That Heal* (p. 60, by L. Koplow, 2002, Teachers College Press. Copyright 2002 by Teachers College, Columbia University).

These 0–5 milestones are foundational to life and in fact even us adults can often find ourselves back within them. Consider a time when you felt emotionally depleted or overwhelmed, did you turn to someone close to you for support? That person was likely your *psychological home base*. What about all of those pictures you keep stored on your phone? Those are adult versions of *transitional objects* (things that represent the important relationships in your life). Or how about when you find yourself in a power struggle with one of your children…yep, you're right back in your 5-year-old self, struggling with *power vs. powerlessness* (and don't worry, we've all been there, more often than we would probably like to admit!)

In this section we are going to take a deeper dive into the five key milestones, along with the Regulation of State milestone, as they all have significant relevance when dealing with both behavior management and digital disruption.

1. Attachment

The readiness to go to specific adults for safety, comfort, and experiences of reciprocal joy.

When we parent our children with **nurture**, reciprocity and **attunement**, consistency, and emotional responsiveness, we lay the path for the development of relational security. This means that our kids learn to trust that other adults will take care of their needs, that the world is safe, and that they themselves are worthy of love. A secure attachment style is crucial for social and emotional well-being throughout life leading to trusting and healthy adult relationships.

Another bonus? When kids are securely attached, the Downstairs Brain switches off its survival alarm system leaving them free to explore and relate with the world around them.

Real Talk:

"I never realized that all of those little moments, those hundreds of little moments could, and would, result in something as significant as a secure attachment style. The gentle touching of tiny fingers and toes, the smiling at and singing to, the acceptance of smiles, cries, fears and discomfort, the use of facial expressions to mimic my infant's feelings, the persistent rocking, swaying, holding, and just being with. All of these repeated connections ultimately led to my toddler's ability to embrace new family members, trust daycare providers, interact with friends, and explore her world, all while knowing she is loved beyond measure."

When it comes to attachment, the predictable, nurturing, unconditional responsiveness of a caregiver is the key to security, which in time allows your child to accept **structure** and **nurture** from you, feel emotionally safe when **sitting in the yuck**, and know that **natural and logical consequences** are not a reflection of them as a person but rather just their behavior choice.

2. Regulation of State

The ability to manage one's internal world to function effectively in the external world.

A child's ability to regulate (or more simply put, manage) their physical and emotional state begins to develop right away, through the act of co-regulation. Co-regulation is the mutual process where individuals influence each other's emotional and behavioral states, particularly in moments of stress or heightened arousal. It involves a supportive interaction where one person helps another return to a calm, regulated state. It first requires adult self-regulation, meaning you have to be as cool, calm, and collected as possible. Then, your regulated state of calm helps to influence stress hormones in your child, helping them calm down too.

Think back on all those nights spent rocking, soothing, swaying, and pacing the hallways, infant or toddler in your arms (not to mention all the deep breaths you had to tell yourself to take). That's co-regulation in action.

These experiences must continue until consistent self-regulation skills have been established, which, by the way, can take years, and even as adults we will struggle to self-regulate sometimes!

Brain Bite: A child *cannot* learn to self-regulate independently: Children can *only* develop their capacity to control physical and emotional energy through this scientifically based sharing and receiving of calm, in the context of a healthy, safe relationship.

Real Talk

"My 2-year-old came home from daycare the other day and she was so cranky! I had to get dinner ready, and honestly, I was feeling really frustrated. I kept trying to encourage her to play with her toys independently but she kept getting mad and would start crying and I would have to stop what I was doing to respond to her. It was hard; I was definitely impatient to begin with. But once I realized that her body and brain really needed me, instead of things, I took a deep breath and told myself that dinner could wait. We sat and snuggled quietly for a bit and looked at some books. After a while, she was much more content with looking at her books by herself and I was able to get started on dinner again."

Don't panic! You don't have to be a complete "Zen" parent, but it is helpful to be mindful of your emotions. The Child Mind Institute highlights this perfectly stating that *"Co-regulation does not mean pretending to exist in a state of calm all the time or never getting angry. It means actively managing your own emotions to help kids learn to manage theirs"* (Hagen 2024).

Making sure you are regulated should always be the first thing you do when your child is in a dysregulated state. We recommend taking deep breaths as a tool for getting regulated.

Co-regulation experiences set the stage for a child's understanding of their **window of tolerance** and ability to **independently regulate** in the future, and are highly dependent on the adult's **attunement** to the child's needs.

3. Psychological Home Base

A sense of belonging in which self-identity is tied to a particular person.

In case you haven't realized it yet, you are your child's headquarters for everything safe, reliable, and free of judgment, and the place they can go to emotionally refuel through the good, bad, and the ugly. You're what we refer to in the child-development field as a "psychological home base." Sometimes this means we have to **sit in the yuck** with them, set **structure** (balanced with **nurture**) and use **verbal responses** to validate experiences.

Without a psychological home base, a child may lose trust in the adults around them to keep them safe, support them without judgment, and be emotionally available for them: They will no longer go to adults during times of need. Instead, they may rely on other options, including digital sources such as tablets, phones, and other devices, which can often lead to more problems, not solutions.

Real Talk

"I was watching my 3-year-old granddaughter at the playground the other day. I noticed her interacting with another child of a similar age. They were playing with a ball. The other child took the ball and started running away with it. I noticed my child try to get it back and begin to get upset. I saw her turn and look for me and then run to me while crying. For a few moments I just held her and soothed her. When her crying subsided, I voiced that it can make us sad when someone has something we would also like or when others don't share, and that I was sorry that happened. I encouraged her to either ask the other child to play with her, or choose another activity. She smiled and ran to swing instead."

Don't worry, being a psychological home base doesn't mean having all the answers or always understanding what your child is going through or feeling; it's simply welcoming them however they are showing up with open arms and listening, and non-judgmental ears. By doing so, you are setting the scene for all the good, bad, and ugly times in their future that will undoubtably show up (including when you have to increase **structure** around electronics or implement **natural and logical consequences** around social media and screen time use).

4. Object Constancy

The ability to maintain a stable emotional bond with another person, even when they are physically absent, or when there are negative experiences within the relationship.

"Peek-a-boo? What's that?" said no parent ever.

Yep, that game you played, A LOT, has real scientific purpose.

Brain Bite: Peek-a-boo offers more than entertainment. It helps your infant learn that people and objects continue to exist even if they can't see them and that they come back when they go away (the scientific term for this is *object permanence*). The constant repetition of this activity and others like it (e.g., hide-and-seek) builds strong brain connections that reiterate this message.

Object constancy builds upon object permanence. Simply put, object constancy means a child understands that even if Mommy looks angry or is away on a trip, she's still the same Mommy and their bond hasn't changed.

Object constancy requires a secure attachment style in order to develop; it is built on trust and the ability to hold an internal positive view of the self. For example, our child might see that we are upset about a behavior they exhibited, but they know that our relationship with them is still solid and that we love them. They understand that they are a good person and worthy of love, even if they sometimes behave in ways that aren't helpful, safe, or kind.

Real Talk

"I remember when Dante was 4. I made a quick trip to the grocery store and put him down to grab a shopping cart to put him in. In that split second, he ran away from me, almost running into the road. I grabbed him and yelled, "No!" My face was mad and my voice was loud. He started crying. I hugged him close and gently told him that I was sorry and that I loved him and just wanted to keep him safe, reminding him that I was still Mama, even if I had looked and sounded a bit different in that moment."

In our field of work, we call these types of instances "ruptures" and "repairs." In the scenario above, Dante's mother's behavior (raising her voice) and change in facial expression created a minor rupture in their relationship—a moment in which Dante was unsure if his mother still loved him and if their relationship was stable. By hugging him and using soothing tones, Dante's mother was able to repair the rupture and deliver the message to Dante (and his brain) that she did indeed love him still and their relationship was unchanged.

This isn't a time to worry. Minor ruptures and repairs are normal and a healthy way for our kids to learn object constancy. It's important to note that developmental harm will only occur if ruptures are frequent and associated with pain or fear, and repairs are inconsistent or non-existent.

For a powerful look at this process, we highly recommend watching Ed Tronik's Still Face Experiment video (Tronik, 2009). Not only will you get to see the foundations of object permanence in action, as well as the experience of rupture and repair but you can also begin to understand how damaging it can be when we as adults become too distracted by our own screens.

5. Mastery

The intrinsic drive and persistent effort to learn, explore, and control the environment, often without external rewards.

Mastery relates to the child's ability to think, reason, and process at a much higher level of complexity. It is a fundamental aspect of development, indicating a sense of competence and the ability to tackle challenges. This drive is crucial for acquiring new skills, **problem-solving** and ultimately, resilience.

Mastery is about taking risks and sometimes failing, which can lead to external dysregulation, such as crying, screaming, throwing things, or shutting down. It's important to remember that these behaviors are driven by internal feelings the child might be experiencing while learning new skills (i.e., frustration, embarrassment, anger, or sadness). Mastery becomes apparent when children attempt tasks multiple times, experience failure, but finally succeed and feel proud of their accomplishment.

This process requires:

- knowledge that they can turn to the adults around them when things get hard for help as well as experiences of reciprocal joy when they experience success (a secure attachment style);
- basic self-regulation skills and, more importantly, an adult who can co-regulate when dysregulation occurs (regulation of state);
- reminders that they can be successful, even when they fail, giving them the courage to try again (a psychological home base); and
- an understanding that mistakes will happen during learning, I may fail and feel like I've disappointed people, but I am still loved (object constancy).

As you can see above, by the time a child reaches the age of 4 or 5, it becomes very apparent how important it is for earlier milestones to be well-established in order for them to continue to develop healthy social and emotional skills.

Real Talk

"When Skylar turned 5, she began learning how to ride a bicycle without training wheels. It was a fun, and frustrating, experience! It took quite a few evenings of us helping her, practicing, falling off,

and a few tears before she was able to do it by herself. At one point I even remember her telling us that she was sorry that she couldn't do it. Hugs and patience were our go-to supports! I'm so glad she finally achieved it; she was so happy, we went to get ice cream to celebrate."

Quick Tip(s): Sometimes it can be hard to watch our children fail at something and experience the hard feelings that go along with it—we just want to take the hurt away! We have to remember that trying and failing sets us up to know that when things get tough, we have the ability, and courage, to try again. **Sitting in the yuck** is vital. A common childhood rite of passage such as learning how to ride a bike has so much more depth to it than we might realize!

6. Power vs. Powerlessness

Power, in the context of child development, is the feeling of having control over one's own actions, choices, and environment. It encompasses a child's developing sense of self-efficacy, autonomy, and capacity to make decisions and influence outcomes.

Between the ages of 4 and 5 children want to be needed, feel important, and have more independence. They express this by frequently asserting themselves. There's nothing quite like that feeling you get when your child tells you, "No!" for the first time and really understands the meaning behind it. Maybe it went something like this… 😬 🤣 😭

In order for our kids to understand power and powerlessness, we must create opportunities for them to make choices, take safe risks, and feel powerful. By doing so, we are building their willingness to explore and take risks independently, which is crucial for later learning experiences. It can be useful to remember the **importance of play** as we help children learn this skill.

🗨 Real Talk

Here's Josiah, who's 5, and his dad. It's almost bedtime and this is often a hard transition for Josiah (like most other 5-year-olds out there). Josiah's dad uses what he knows about power and powerlessness to help support Josiah at this time:

Dad: Hey, buddy, it's getting close to bedtime. We have time for two books or one show before we go upstairs. What would you like to choose?

Josiah: A show! I want to watch a show!

Okay, do you want to choose Paw Patrol *or* Bluey*?*

Paw Patrol!

Great choice! When we are done watching we have to go upstairs. Do you want to race or I can carry you?

Race! Race!

Okay, another good choice. I'll say, "Ready, set, go" when the show is over.

(Eight minutes later…)

Okay, buddy, are you ready to race me?

No! One more show!

I know those shows are fun. It's bedtime, remember? And you chose to race me upstairs! Get ready, otherwise I might win!

(Jump up, and get ready to race. Be playful and excited.)

I said no!

Phew, it seems like you're having a hard time right now. You can choose to race me upstairs for a bedtime story or you can choose to go to bed without a story.

No!

It seems like you need me to choose for you.

No, let's race!

Okay, great choice. Ready. Set. Go!

During this interaction, Josiah's father successfully turned bedtime struggles into moments of connection by utilizing **structure** combined with choices, while remembering the **importance of play**.

As they progress through this milestone, children need adults who are present with them through the emotions and behaviors that occur while they are learning about power and powerlessness—sometimes this means **sitting in the yuck** during uncomfortable, frustrating, and sometimes hurtful moments.

Quick Tip(s): If it's safe to do so, bringing **play** into your interactions with your child will almost always help avoid or reduce dysregulation, as it shifts their attention away from the difficult transition they are being asked to take part in. Activities such as pretending, humor, singing, and make-believe move the brain away from stress and toward connection. Side note: **Play** doesn't just work with little kids—big kids respond positively to fun too, it just looks and sounds different with them.

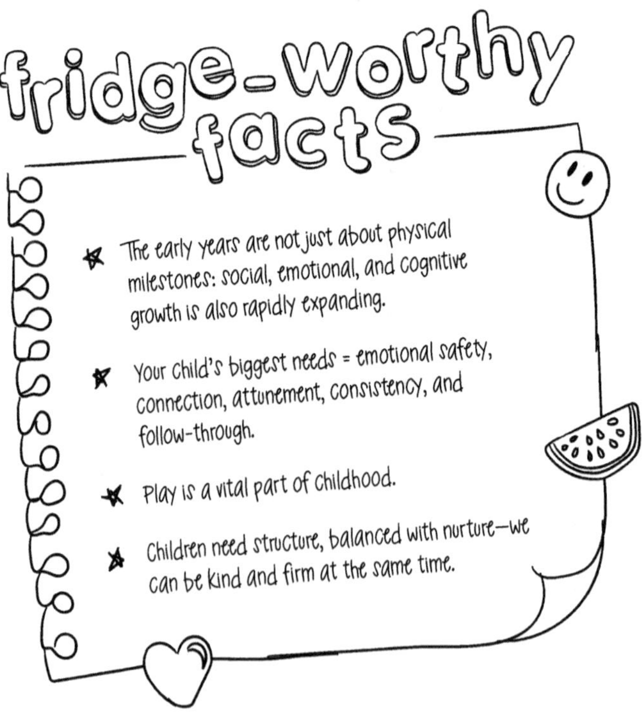

DEVELOPMENT IN THE CHILDHOOD YEARS (AGE 5–AGE 10)

"From Imaginations to Independence"

Between the ages of 5 and 10, our kids begin to develop improved cognitive abilities like longer attention spans, logical reasoning, complex **problem-solving**, enhanced memory skills, and a deeper understanding of perspectives other than their own.

Not only that, but we see significant growth in language skills and the ability to process information more efficiently. (This might explain why our kids also become much better at arguing with us!)

If you suddenly find yourself chopped liver, don't take it personally. This is a time when children are becoming much more independent, are learning more about their place in the world, starting to think about the future, and developing more significant friendships and peer groups. Learning happens through experiences such as reading, games, sports, and diverse activities.

During this time, supportive relationships with family members and peers play a vital role in healthy brain development. Your use of **verbal responses** to help children deepen their understanding of how their brains and bodies work together will be vital.

The social and emotional skills that children learn through these ages are built directly upon the early milestones. For example, understanding that they can feel two emotions at the same time ("I like John, but I hate how he talks to me") is an extension of object constancy, and viewing themselves based on how they perform in school, their capacity to make friends, or their physical appearance will all be impacted by their inner beliefs about the self, which are founded in their attachment relationships.

Brain Bite: The Upstairs Brain is still under construction and is busy making connections to support higher levels of cognitive functioning. Children at this stage will impress you some days with their emotional maturity but may get upset over something trivial like a cereal bowl on other days.

Therefore, throughout these ages, children need adults who are understanding of the fact that some days they will "get it" and other days they won't. We may have to **sit in the yuck** with them and help them cope with **natural and logical consequences**.

Adults who can provide a good balance of **structure and nurture** and who can help them navigate the complexities of **problem-solving**, peer relationships, and their own belief system with empathy and patience are a must.

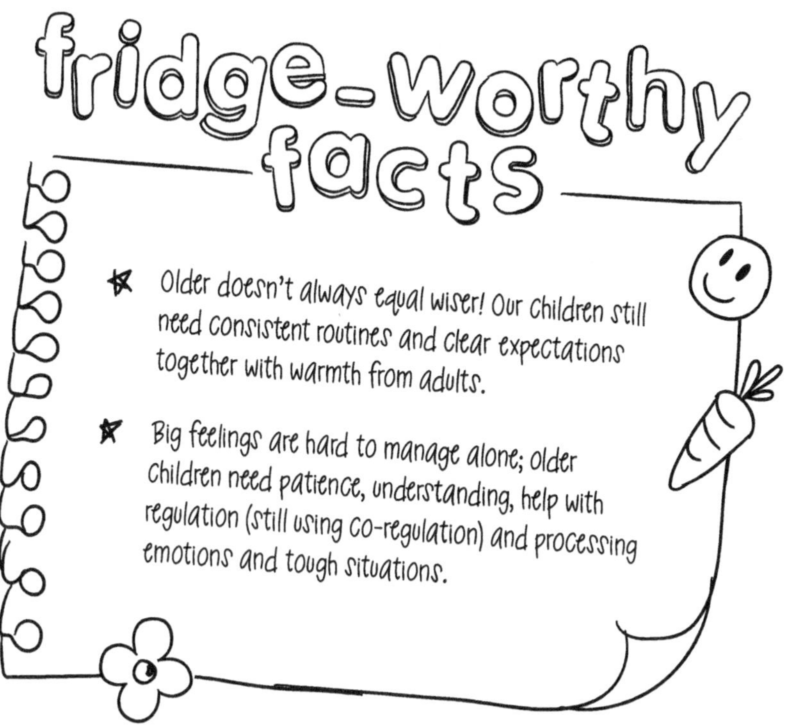

DEVELOPMENT IN THE EARLY TEEN YEARS (AGE 10–AGE 14)

"Renovation Underway—Expect Loud Noises"

If you've lived with a tween, or you're raising one right now, I salute you. We often joked in our house that we wouldn't know which version of our child would walk down the stairs each morning. Between mood swings, body changes, and peer-group drama, early adolescence is not for the faint of heart.

Parenting a tween is hard. Being a tween is hard.

🧠**Brain Bite:** The changes that take place during the period of adolescence have been described as a "fundamental reorganization of the brain" (Konrad et al., (2013), p.430), and it's important to remember that your young teen's brain is still under development. Imbalances between the different systems in the brain are likely to blame for typical adolescent behaviors such as risk-taking, poor emotion regulation, and impulsivity.

To put it simply, your tween's brain is undergoing a major renovation. And the construction workers disagree, a lot.

Real Talk

"Having a tween is like living with a small, unpredictable roommate who alternates between needing a cuddle, slamming doors, and delivering life advice like a tiny philosopher on too much sugar."

Adolescence is a time driven by intense emotions (tell you something you don't know, right?) and the desire for new experiences, especially those outside of the norm. Learning happens through experiences with peers, at school, in teams, and during social activities, as well as through hobbies such as art or music. It is a time when our children are becoming more independent while also exploring stronger connections with friends.

Though it might not seem like it sometimes, I promise your young teen's brain *is* starting to get better at more cognitive processes, including **abstract thinking** and logical reasoning, as well as social and emotional skills such as emotion processing, self-control, and understanding social cues.

Much like the childhood years, social and emotional skills that young teens learn between the ages of 10 to 14 are built directly upon earlier milestones. For example, the ability to think in terms of what might be true, rather than just what they see as true, as well as the capability to more accurately interpret others' emotions is an enhancement of object constancy. Furthermore, the continued development of self-identity, opinions, and values is strongly tied to their attachment style, feelings of power vs. powerlessness, and the knowledge that a psychological home base exists in their life.

Despite the fact that sometimes their actions will make you want to run for the hills, young teens *need* adults in their life who can walk alongside them, helping them take healthy risks, make difficult decisions, giving them permission to make mistakes, setting clear expectations, and enforcing **natural and logical consequences**. Caregivers who are able to **sit in the yuck** with their young teens and help them express and manage their emotions in healthy ways using **attunement** and **verbal responses** will be better equipped to support their children during this roller-coaster ride of growth. It's important to remember that even though learning to navigate independence is a big part of this developmental phase, our young teens still need boundaries and limitations in the form of **structure**, and love and patience in the form of **nurture**.

AN IMPORTANT DIGITAL DISRUPTION CONSIDERATION FOR THE EARLY TEEN YEARS

Brain Bite: During the adolescence renovation process, unused connections in the thinking and processing parts of the brain are "pruned away," giving meaning to the phrase "use it or lose it." In simpler terms, weak connections in the brain will be lost, and strong connections will remain; therefore, how your young teen spends their time will be crucial to brain development.

Real Talk

Ari's 13. Here, her mom describes some of the changes she has seen take place as Ari's choice of activity changed:

"Ari used to play soccer, from the ages of 6 to 11. She was great at it and she used to be all smiles after her practices and games. But over the last few years, she's started playing video games with her friends a lot and doesn't have any interest in playing soccer—or anything outside anymore. Yesterday we were out in the yard and I encouraged her to run around a bit and kick the soccer ball around with me. I noticed the joy I used to see in her come back when we were out there, but unfortunately her skills weren't quite as good as they used to be."

Ari's mom's experience is a great example of how digital media can impact our tweens during this important period of development. The increase in use of video games meant that Ari wasn't using her soccer skills anymore, leading to both her interest and skills diminishing.

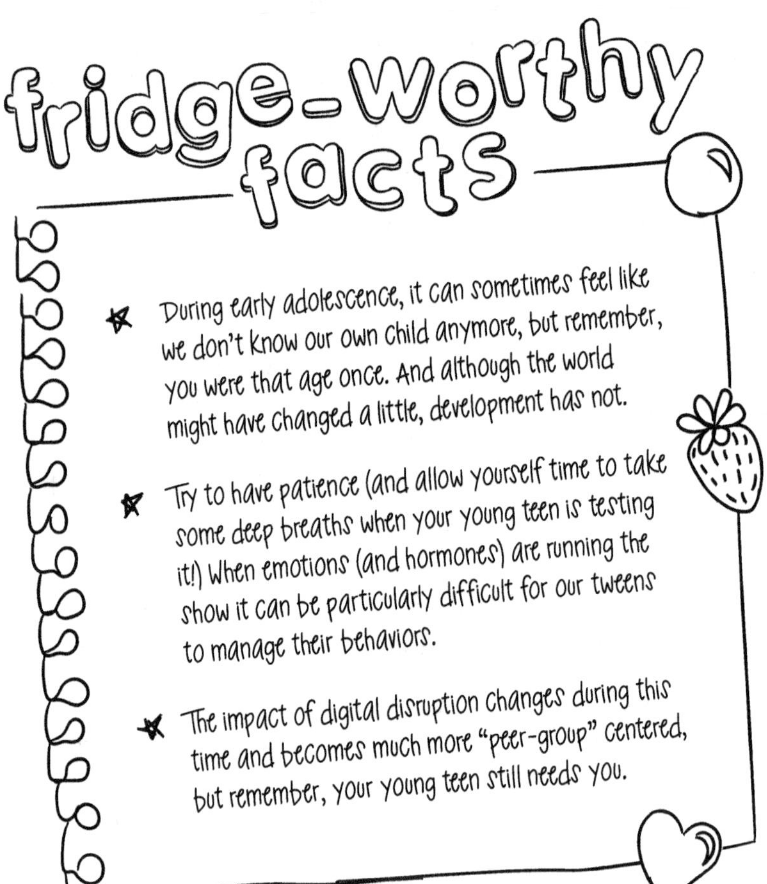

DEVELOPMENT IN THE LATE TEEN YEARS (AGE 14–AGE 18)

"Almost Grown, Still Growing"

They might sound like adults. They might look like adults. But their brains? Still under construction.

While the renovation process might have slowed down a little, there is still significant brain development happening during the later teen years. And even though our children now look like fully functioning adults (and believe they are), the brain's construction won't be finalized until they are in their mid-to-late twenties.

So, if some days your teenager is able to hold a well-rounded discussion with you about the state of the economy but other days they meltdown quicker than ice cream in the sun, all is well. (Regarding their brain, that is. Your sanity? Well, that's up for debate.)

Brain Bite: Older teens definitely have more advanced Upstairs Brain skills such as decision-making, moral reasoning, and complex **problem-solving**; however, oftentimes, strong emotions are still in the driver's seat, causing impulsivity and an inability to really consider the potential consequences of their actions.

Learning continues to happen through experiences with peers, during social activities and in more formal settings such as at school or a part-time job.

More challenging, and sensitive, topics will come up for you and your older teen—life choices, sexual experiences, access to drugs and alcohol, inappropriate media content—and it's important to remember that older teens can't consistently apply advanced thinking skills because strong emotions continue to drive decisions where impulses come into play.

Older teens still need adults who can empathetically support their use of **abstract thinking** and **problem-solving** skills as they make decisions surrounding challenging scenarios that might significantly impact their futures. They might also need a trusted parent or caregiver to **sit in the yuck** with them when they make mistakes and help them create **structure** for themselves to avoid making more. And while they might lead you to believe otherwise, they still need you for **nurture** and experiences of **play**—even though these probably look a little different now.

A FINAL WORD ON DEVELOPMENT

Real Talk

"I wish I had had this information when I was raising my kids, and that was even before all the digital disruption happening nowadays."

We know this probably seems like a lot, and you're right, it is. You don't have to know or remember everything. We just encourage you to understand a little bit at a time, and use what works for you.

Hear us when we say, parenting with your child's brain in mind doesn't mean that everything will be perfect and that you'll never raise your voice or lose your cool. It just means that you might be a little more prepared, a bit less shell shocked, and a whole lot more capable of knowing what to do—and that makes all the difference.

And trust us: The more grace you extend to their growth, the more you'll give to yourself.

You've got this!

Chapter 3

WHEN SCREENS TAKE OVER

When Reality Becomes Virtual Reality

Let's be honest - screen time can often feel like a lifesaver, serving as a critical resource for parents during challenging moments. The toddler is melting down? Hello, Baby Shark video. Need five minutes to make dinner? Cue the cartoons. Stuck in a waiting room? The phone saves the day.

Here's the thing, screens in small doses aren't the enemy. Sometimes they can even turn out to be beneficial rather than harmful. But when they become your default and quick fixes are the daily norm, quiet changes start to occur in our children's brains, bodies and behavior.

A STORY THAT MIGHT FEEL FAMILIAR

Jasmine is 8. She used to spend hours outside playing with neighborhood friends. Now, most afternoons are spent watching shows on the T.V., playing games on her tablet, doing homework on her laptop, or battling her brother at video games. Her parents noticed she was quicker to anger, found transitions hard to accept, struggled with friendships, and hated being "bored".

Screens had taken over her time, her brain, and her social life, and her wellbeing was paying the price.

We call this process "Digital Disruption" - not because technology is bad, but because too much tech can get in the way of real-world interactions, causing disruption to development, resulting in potentially long-lasting consequences and negatively affecting kids' overall wellbeing.

> Digital Disruption occurs when screens interfere with real-world interactions, disrupting development and potentially harming kids' long-term wellbeing.

In this chapter, we are going to take a deeper dive into the impact that frequent, intensive technology use, combined with a lack of connection, and correction, can have.

Digital Disruption in the Early Years (AGES 0-5)

Foundation Shaken: Little Humans, Big Impact

In the early years, our kids' brains and bodies are undergoing some of the most significant changes they will ever go through. Learning happens in the context of a relationship—how to turn off their survival mode, how to communicate, how to manage emotions and how to manage their bodies. This means that when digital disruption takes over there is a very big, and very real, long-lasting impact on social, emotional and even physical development.

Digital Disruption #1

Screens replace relationships.

Little ones are experiencing less reciprocal (back and forth) interactions, emotional responsiveness and connection with safe, predictable adults. This leads to less trust in adults to meet their needs and less relational security (disrupted attachment)—"I'm not sure if people care about me, or if I am loved or valued".

Digital Disruption #2

Screens become the go-to soothing tool.

There are fewer opportunities to develop self-regulation skills through co-regulation. Instead, children are being given bright lights, sounds and screens when they are dysregulated (having a hard time). This leads to young children who cannot regulate their bodies or their emotions and who struggle with transitions.

Digital Disruption #3

Screens replace reality.

Children face less real-world experiences of challenge resulting in less experiences with the feeling of mastery. This can lead them to question their ability to make things happen when they need them to happen.

Digital Disruption #4

Screens replace peers.

Children are experiencing fewer social interactions and less opportunities for play with others, leading to an inability to tolerate and understand feelings of power and powerlessness, difficulty recognizing emotions of others, and understanding and responding to social cues. In the long term, this can cause challenges with the development of higher cognitive skills such as empathy, reasoning and **abstract thinking**.

Digital Disruption in the Childhood Years

(AGES 5-10)

Missing the Moment, Playtime Replaced, Skills Delayed

In the childhood years, kids' brains are still developing, and higher-level social and emotional skills are being built and honed through play, education and social experiences. Digital disruption during these years can have a huge impact on the formation of vital competencies that can result in life-long challenges in numerous areas such as relationships, physical health and education.

Digital Disruption #1

Screens replace real-life connection.

When screens replace real-world interactions it can be hard for children to develop the skills needed to understand social cues, etiquette and other people's perspectives. This can lead to struggles making friends, challenges with family relationships and difficulties expressing needs appropriately in environments such as school or the home.

Digital Disruption #2

The "odd occasion" becomes the obsessive norm.

The lights, music, speed and colors used on screens and in technology aren't there by chance. They are specifically designed to trigger our brains to release feel-good chemicals that keep us wanting more. This makes it incredibly hard for our children to find as much joy in other activities that don't give their brain as much of a boost. This process, unfortunately, can lead to our kids truly becoming "addicted" to screens.

Digital Disruption #3

Interests, activities and vital tasks become second choices.

When screen time takes over, there is little time for other interests and activities such as sports, music, and homework. This can create challenges with social skills, physical skills and physical health and reduced academic progress.

Screen-time obsession also means less downtime, giving kids fewer opportunities to experience boredom, or other challenging emotions. This limits their ability to develop appropriate coping skills to manage uncomfortable feelings and can result in poor self-regulation skills, an issue which often causes problems in other environments such as school or social settings.

Digital Disruption #4

Social media increases exposure to negative content.

If we allow our children to have access to social platforms from a young age, we risk exposing them to negative content such as unhealthy role models or cyber bullies. This can have a significant impact on a child's mental health including self-esteem, confidence, and self-worth, and can cause increased worry, fear, anger or confusion—this is especially true for the older child whose brain has not yet developed higher level cognitive functioning.

Digital Disruption in the Early Teen Years (AGES 10-14)

Emotional Earthquake: The Tween Tech Trap

In the early teen years, the brain starts its process of renovating and reorganizing, and growing, once again. Big changes are happening, some of which you can easily see (such as that giant growth spurt), and others are much harder to spot (such as strengthening of neural connections leading to improved cognitive ability). Regardless, digital disruption during the tween years has the potential to interrupt these processes in a big way. This can compound challenges in already tricky areas such as decision making, peer relationships, and learning.

Digital Disruption #1

Screen content increases, supervision decreases.

As kids gain independence, supervision naturally declines—especially around screens. This gives tweens greater access to harmful content, unhealthy peer influences, and cyberbullying, all of which can impact mental health and increase risk for self-harm and suicide. Algorithms can expose teens to dangerous content even without searching for it. Repeated exposure may normalize these behaviors, particularly for those lacking support or coping skills, and can trigger vulnerable teens to engage in risky behavior (George, 2019).

If your child is experiencing feelings of suicide or self-harm and you need support, please go to page 307 to access our list of recommended resources. There are organizations out there that can provide help to both you and your child. You don't have to do this alone.

Digital Disruption #2

Connecting through the screen, isolating in person.

Young teens want to spend their time on their screens, whether that's a phone, laptop or video game. This increase in isolation often means less time connected with caring adults in the real world. As a result, the amount of support our kids seek out or receive when they are experiencing difficult emotions or experiences decreases. In fact, oftentimes, the screen becomes their coping method. Over time, this can create significant challenges for their emotional development and can result in feelings of depression, loneliness and low self-worth.

Digital Disruption #3

More screen time = less skill time.

Too much screen time in the early teen years means that there's less time (and focus) on other important activities such as homework, family time or chores. As a result, our kids can have difficulty understanding and managing responsibilities. At a time when their brains are pruning away less-used information, these vital skills may be lost or not even developed, making later life much more challenging for them.

Digital Disruption #4

Tech takes over taking care of self.

As kids get older their access to physical activity at school decreases. Add to this an increase in screen time (and therefore less time for play, sports or recreation) and we've got teens who run a real risk of obesity and other physical health problems.

Furthermore, the addictive pull of the bright light, sounds and access to friends 24/7 mean that sleep is often sacrificed for more exciting activities. Restorative sleep is vital in the early teen years and a lack of it can cause a significant impact on mood, cognitive function and ability to manage tasks and responsibilities.

Digital Disruption in the Late Teen Years

[AGES 14-18]

Virtual Risks, Real Consequences

The late teen years bring greater independence, but also greater risks. While our teens' brains are indeed more developed and skills such as problem-solving, abstract thinking and decision making are considerably improved, social, cultural and academic pressure can all impact the capacity to use these abilities successfully. Digital disruption during these years can impact future potential and opportunities in ways that our teens likely haven't even considered.

Digital Disruption #1

Better brains + more independence = more access + more risk.

Increased accessibility to harmful or inappropriate content (due to our teens' intellectual ability to find such media) can increase their risk of exposure to dangerous trends and explicit content, including pornography, drug-use, extreme weight-loss and other unhealthy behaviors. Over time and combined with increased independence, this can lead to numbness or acceptance of such behaviors as normal. This then increases risk for substance misuse, mental health and eating disorders, gambling, pornography and sexualized behavior, which in turn impacts their view and understanding of healthy relationships.

Digital Disruption #2

Greater risk, greater consequence.

As their use of technology becomes increasingly independent our older teens run the risk of making decisions that can pose a great risk to their future depending on the extent of their behaviors. For example, online gambling, engaging in sexualized activity with minors (even if it's unintentional) or drug and alcohol use that becomes public knowledge can all have legal consequences that may impact future college or career opportunities.

Digital Disruption #3

Increased online connection, real-world disconnection.

Like our younger teens, older teens can become increasingly engrossed, and dependent upon, their online world. By disconnecting from the real-world, our teens now face a lack of access to healthy outlets, healthy peers and healthy activities. This means that when they experience difficult or challenging emotions, they are more likely to turn to their online world for support—often resulting in unhelpful, unhealthy and potentially harmful, coping strategies.

Digital Disruption #4

Real risk, fake people.

More time spent online means higher chances of our teens being contacted by online predators—individuals who pose as peers to build trust to trick teens into sharing explicit photos or engage in inappropriate conversations and then threaten to publicize them unless money is paid or more content is produced.

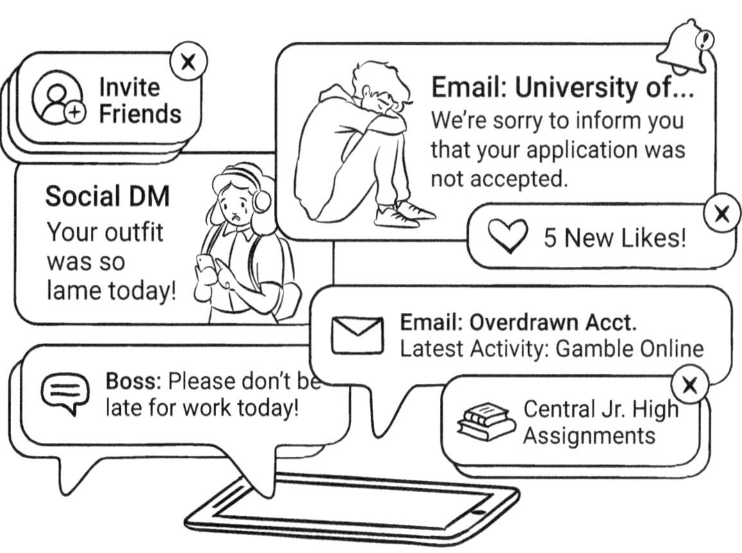

Although this information can be alarming, there are steps you can take to mitigate the risks, no matter what your child's age is. There is no need to discard screens entirely. You just need to rebalance the scales.

That starts with awareness.
With choosing play over pixels.
Connection over correction.
Eye contact over eye strain.

THE OUTCOME FOR JASMINE:

Jasmine's parents recognized that her use of screens had taken over other priorities in her life. When they gently reduced the amount of screen time Jasmine had, things got worse before they got better—there were tantrums, tears and other withdrawal symptoms. But slowly, new patterns emerged. They created a family screen plan, prioritized evening walks and board games and encouraged Jasmine to write to her cousin instead of texting.

Two months later, Jasmine was smiling more, didn't complain so much of boredom, and even played outside with her neighbors again.

This isn't a miracle story—it's a reflection of how shifting the balance can begin to restore the functionality of the brain and promote healthy development.

fridge-worthy facts
Screens aren't evil, but overuse rewires the brain and displaces essential real-life development.
Screen time needs limits, social media use needs monitoring, and real-world connections need intentional protection.
Digital disruption is different across the ages, but kids still need caregivers, whether they are 5 or 17.

Chapter 4

THE DTAP® SOLUTION

Co-op Mode Unlocked: Science + Relationships = Being With, Not Against

Some parenting books give you checklists. Others give you charts. DTAP® gives you something deeper: a way to see your child through the lens of connection, regulation, and their developmental needs instead of merely responding to their behavior.

Because it's actually about understanding what's beneath the behavior. And DTAP® helps you look there first.

DTAP® stands for Developmental Trauma and Attachment Program, and it's at the heart of how we help kids at Chaddock heal, grow, and thrive. But don't let the fancy name fool you; it's built on something ancient and human—relationships.

On the following page you will see a simplified picture of the DTAP® Pyramid. Most behavior-based approaches focus on the outside—what the child is doing. Our pyramid helps you understand what's going on inside—which part of the brain your child is operating from—and what response will connect most with them at that time (*and* be most helpful in managing the behaviors on the outside).

Because here's the thing: If a child is in their Downstairs Brain (survival mode), they can't reason. They can't reflect. And they sure can't "make better choices"—those skills all require them to be in their Upstairs Brain. They need someone who is cool, calm, and collected (i.e., in *their* Upstairs Brain) to greet them downstairs, help them to regulate, and then support them as they climb the stairs back to logic and sense, as well as more helpful behaviors.

Next-Level Knowledge: Look back at chapter 1 if you need a reminder about the different parts of the brain.

The DTAP® model is there to help you:

- discover which brain region your child is functioning in;
- choose responses that calm, connect and guide; and
- rebuild trust, emotional safety, and regulation.

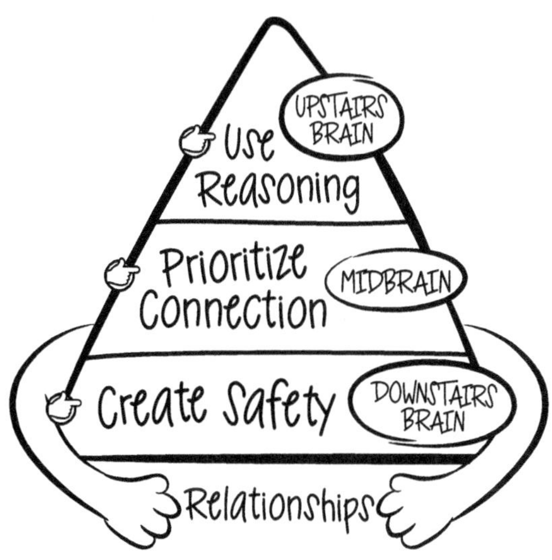

WHY RELATIONSHIPS COME FIRST

The most important thing to remember is that relationships are the thread that runs through the entire DTAP® treatment approach: We *always* prioritize our relationship with the child or young person with whom we are working.

This means asking yourself before, during, or after any interaction or when using any strategy, *"Am I maintaining the relationship with my child, or hurting it?"* This approach does not mean we let our kids "get away" with problematic behaviors nor does it mean letting go of **structure** and limits (you will hear us reiterate this consistently in many areas of this book). Instead, it means:

- staying connected during tough moments,
- repairing ruptures when things go sideways, and

- prioritizing emotional safety alongside behavior guidance,

all while maintaining boundaries and expectations.

While DTAP®was originally a framework used to help kids heal from developmental trauma and disruptions in their caregiving relationships, we soon realized that our approach and the concepts we identified within it could benefit *all* kids.

This means that once you understand the core concepts, you can use them as practical strategies to address the specific issues of digital disruption, such as excessive screen time, meltdowns when electronics are removed, or inappropriate internet usage.

Real Talk

"I thought I needed more consequences. What I actually needed was more connection."

"DTAP® gave me language for what I was feeling and what my daughter was experiencing. Now I pause before reacting."

"My son didn't need more rules. He needed someone to understand why he was struggling to follow them in the first place."

Quick Tip(s): Before interacting with your child (or indeed anyone), ask yourself, *"What part of the brain are they most likely in?"* And, *"What part of the brain am I in?"* These questions (and the answers) are especially helpful when responding to problematic behavior.

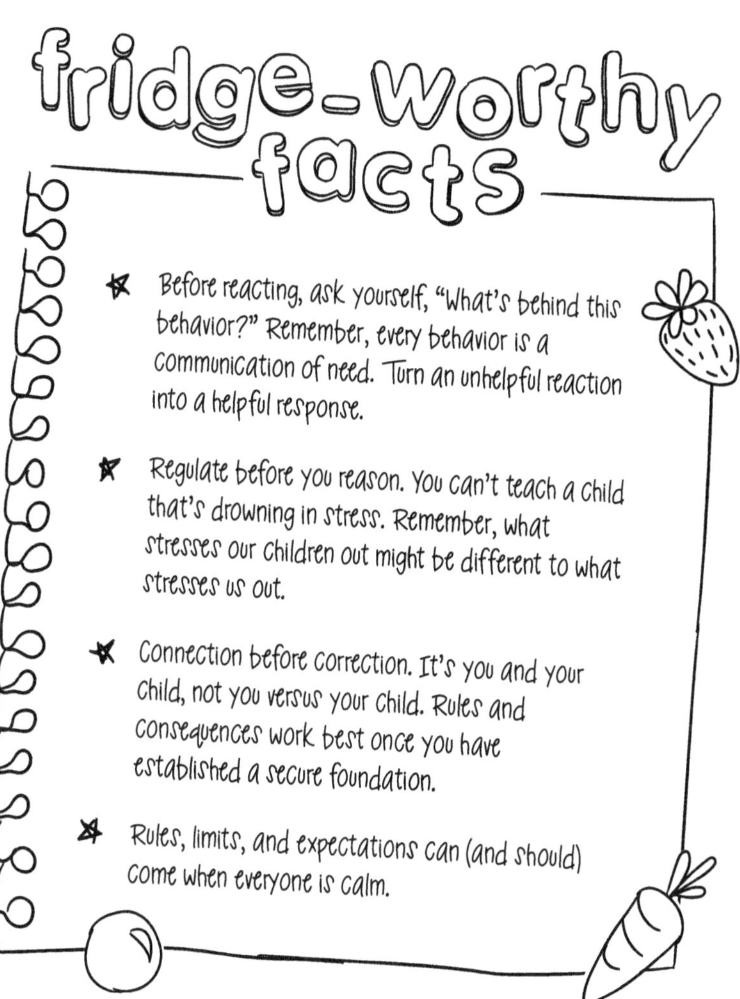

Finally, remember, we aren't looking for perfection—even we professionals get it wrong sometimes. We still react when we mean to connect. The most important thing is that you remember to reconnect when that happens. You're building brain pathways every time you offer a moment of calm, every time you stay instead of walking away, and every time you circle back after things fall apart.

We're human, not perfect. And grace is allowed.

Chapter 5

THE DTAP® CONCEPTS

All the Intel You Need to Complete Your Mission

WELCOME TO THE TOOLBOX

In this chapter, we are going to introduce you to the core DTAP® concepts that drive the way we interact with, and respond to, challenging behaviors that our kids exhibit. These aren't quick fixes; they are deep-rooted, developmentally informed principles. And with a bit of intention and practice, they will help you meet your kids where they are at (i.e., which part of the brain they are in) and avoid prolonged meltdowns and disconnection.

Whether you are dealing with tech tantrums, social media meltdowns, or video game violence, these concepts will help you understand the underlying causes behind such behaviors and will guide your responses, allowing you to maintain emotional connection and promote self-regulation and social/emotional growth.

This chapter has been broken down into three sections based on which part of the brain your child or adolescent might be in—feel free to read all of them, or just the ones that apply to your needs—but remember, the more you understand, the better your ability to respond and reduce digital disruptions.

Downstairs Brain: When Behaviors Are in Charge

1: Attunement

2: Window of Tolerance

3: Structure and Nurture

Midbrain: When Emotions Are Overwhelming

4: The Importance of Play

5. Verbal Responses

6. Sitting in the Yuck

Upstairs Brain: When Thinking Is Possible

7. Problem-Solving and Abstract Thinking

8. Independent Regulation

9. Natural and Logical Consequences

CONCEPT 1
Attunement

"Tune in before you take charge"

Early life milestones that show up here: Attachment, Psychological Home Base, and Regulation of State.

Brain Bite: **Attunement** is especially helpful when your child or adolescent is in their Downstairs Brain and behaviors seem confusing.

Have you ever had to sit in a really long meeting or training and found yourself wishing the person talking would see from your face that you need a break? Or have you been in a situation that was really uncomfortable for you, perhaps a parent/teacher conference where the teacher just kept telling you all the things your child was struggling with and you wished you could just have a moment to share your side of the story?

Attunement is the ability to tune in to your child's emotional state and then respond to it—matching your emotional frequency to your child's. Babies need it to feel safe. Toddlers use it to co-regulate. Teens depend on it to feel seen.

Attunement involves:

- watching facial expressions and body language,
- listening beyond words, and
- being aware of functioning as it relates to sensory, physical, or relational needs.

When we are **attuned**, we are able to help recognize our kids' needs, sometimes before they even know they have one.

Real Talk

Ella, age 4, screamed every time she got dressed. Her foster mom noticed and started narrating the experience for her: *"These socks feel tight. You want to be comfortable. I get that."* The screaming didn't stop overnight, but over time Ella started asking for help instead of screaming.

DIGITAL DISRUPTION!

A kid or teen who is "zoned in" to a screen or digital media is much less likely to be paying attention to important needs that are not being met until it is too late. When they are then asked to stop interacting with the screen or media, these needs suddenly become apparent and can quickly turn into challenging behaviors.

Tuning into core needs and how we can respond:

Need	Example of Need	Digital Disruption Example	How to attune
Physical	Eating, drinking, using the bathroom, sleeping, temperature	The teenager who stays up too late looking at or playing on their phone and cannot get up in time for school.	Think: When was the last time the child had a drink or a snack? Is the child more fidgety, irritable, or unmanageable in hot or cold environments? Is the child more dysregulated at certain times of the day? Are they getting enough sleep?

Need	Example of Need	Digital Disruption Example	How to attune
Emotional	Recognizing, labeling and processing emotions	A 7-year-old who explodes into anger when he loses at his video game, or a 15-year-old who shuts down because a peer said something about her outfit on social media.	Ask: "Help me to understand how you feel about that." "Your face looks sad, are you feeling sad or something else?" "I wonder how it feels when someone says that?" "Wow, that was a lot to take in. Do you feel angry? Or maybe scared or embarrassed?"

Print It, Post It, Pass It On: This is a good chart to stick on your fridge for quick reminders on staying **attuned**.

Need	Example of Need	Digital Disruption Example	How to attune
Sensory	Receiving and responding to environmental information through the senses (sight, sound, touch, taste, smell)	A 4-year-old who spends two hours before bed watching loud videos on a tablet, the large TV is playing a movie, and music is also playing in the next room. The child has a meltdown when they are taken to their dark, quiet room to sleep.	Think: What can I see in this environment that might be overstimulating? What can I hear in this environment? Are there any tastes or smells that might be unpleasant to some but not to others? Would I be uncomfortable if I had to wear that/sit there/touch that?

Need	Example of Need	Digital Disruption Example	How to attune
Movement	Types of movement that help us regulate and stay in, or get back into, our window of tolerance.	A 10-year-old who comes home from school and sits down to play video games for two hours before dinner. By dinnertime, they are extremely grumpy from sitting down for most of the day.	Think: What kind of movement helps my child regulate when they are feeling stressed? Does my child need BIG body movements, such as running and bouncing a ball, or close, low-level movement such as massage to help them calm? When and how much of this type of movement is needed throughout the day? How can I provide it proactively?

Print It, Post It, Pass It On: This is a good chart to stick on your fridge for quick reminders on staying **attuned**.

Brain Bite: Children and adolescents can struggle to recognize their own needs because their Upstairs Brain is not fully developed. That's why it's important for us to recognize and verbalize for them.

Attunement to Stressful Events

As parents or caregivers, it's helpful to stay aware of how certain situations might feel stressful for our kids—even if they don't seem like a big deal to us. Being tuned in to changes in your child's behavior or mood before and during a tough moment can make a big difference in how you support them.

Some digital situations that can be pretty stressful for children or teens include:

- being told no with regards to a tablet, phone, or video game (actually, being told no about anything can be stressful for our kids!);
- transitions* (especially going from something fun—like video games—to something less fun, like chores);
- using, and stopping use of, electronics;
- losing or having limits on screen time; and
- not being allowed on social media apps their friends are using.

Isn't it also hard for us grown-ups to be told no, or have to stop doing something we enjoy and do something less pleasant, or even, dare we say, turn off our cell phones?

*A Deeper Look at Transitions

When we talk about "transitions," we mean those moments when your child has to shift from one activity or environment to another—and that shift can be really tough. These changes often throw kids off emotionally or behaviorally.

🧠**Brain Bite:** Transitions require the brain to refocus—and sometimes to calm down or speed up, depending on what's coming next (Hughes and Baylin, 2012). That mental shift can take a little while, and if it's rushed, it can lead to meltdowns or frustration.

Think of it like this: If someone woke you up suddenly with blaring alarms and started firing off work questions, you'd probably feel frazzled and off your game. Compare that to having a slow morning, coffee in hand, before diving into your day—you'd feel way more in control. It's the same for kids. Even kids who are usually pretty even-keeled need time to adjust when something new pops up, whether it's meeting new people, going to a different place, or just switching up the routine.

Hear us loud and clear when we say, we aren't asking you to let your child or adolescent "get away with" inappropriate or disrespectful behavior when they are asked to stop a preferred activity. Instead, we are suggesting that you offer support, such as choices and reminders to make the transition a little easier for everyone.

Quick Tip(s): The next time you have to ask your child to stop doing something they like and start doing a chore, try an approach like this:

"Pretty soon it's going to be time to put the phone down and empty the dishwasher—I know this isn't your favorite thing, so do you want to do it in five minutes or ten minutes?"

"Ten minutes."

"Okay, great, I'll let you know when you have five minutes left."

"Hey, in five minutes it's going to be time to put the phone down and empty the dishwasher; I'll let you know when you have one-minute left."

"Okay, you have one minute left, then it's time to empty the dishwasher."

"It's time to empty the dishwasher now. Do you want to put the phone on to charge or do you want me to?"

Remember, when it comes to **attunement**, nobody is perfect. And some days you're going to miss the signs—we all do, so try not to worry! When this happens it's helpful to say, *"Oh wow, your behavior was really trying to tell me something and I missed it! Now I know, I can support you better next time you are feeling this way."*

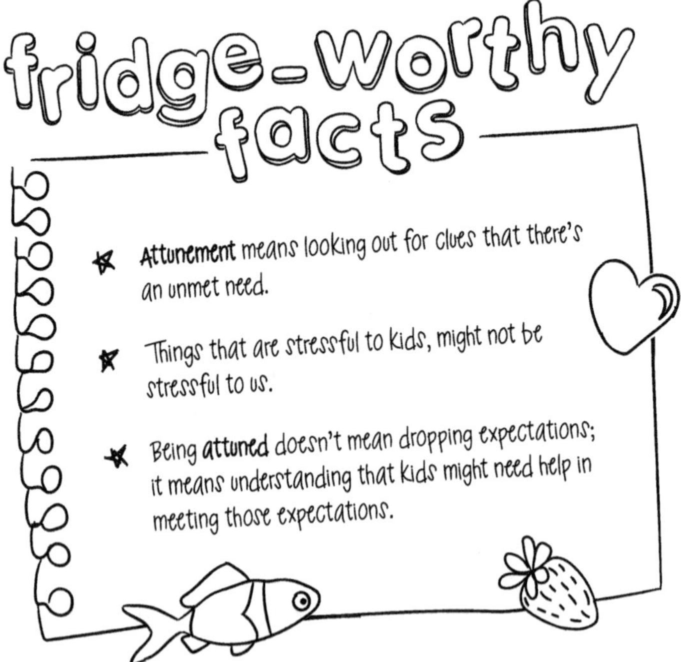

CONCEPT 2
Window of Tolerance

"Behavior gives us a peek inside"

Early life milestones that show up here: Attachment, Psychological Home Base, and Regulation of State

---

Brain Bite: Understanding the **window of tolerance** is especially helpful when your child or adolescent is in their Downstairs Brain and behaviors seem extreme or inappropriate for the time and place.

Imagine being told that you have to sit still for a number of hours and listen to someone speak. Prior to this you have had a snack, a drink of water, and the temperature is comfortable. You are able to sit relatively still for the first couple of hours, but then you begin to get hungry, thirsty, and cold. Your legs start twitching, you move your body to keep warm, and you can't seem to focus on what is being said. When the person sitting next to you asks if you are okay, you snap at them and tell them to mind their own business.

Imagine your child is a snow globe. When life is calm, they can think clearly. But when shaken—by stress, conflict, or even too much screen time—things get cloudy and they get overwhelmed.

Their **window of tolerance** is the zone in which they can manage stress, stay connected, and think clearly. The closer they get to the edge of their **window**, the more intense or frequent, problematic, disruptive, or dysregulated behavior can be observed (Hershler et al. 2021).

When a child or young person is fully dysregulated, that is, in their Downstairs Brain, their behaviors are likely to look, and be, out of place for the situation.

For example, when your toddler has a tantrum in the middle of the grocery store and throws themselves on the floor, this is their way of saying, *"Hey, I'm moving outside of my* **window of tolerance** *into a hyper-aroused state!"* Or if your teenager puts their headphones in, puts their hood up, and stops talking to you, they, too, are moving outside of their **window of tolerance** but into a hypo-aroused state.

And guess what? That example at the beginning of this section? That's an adult moving out of *their* **window of tolerance**.

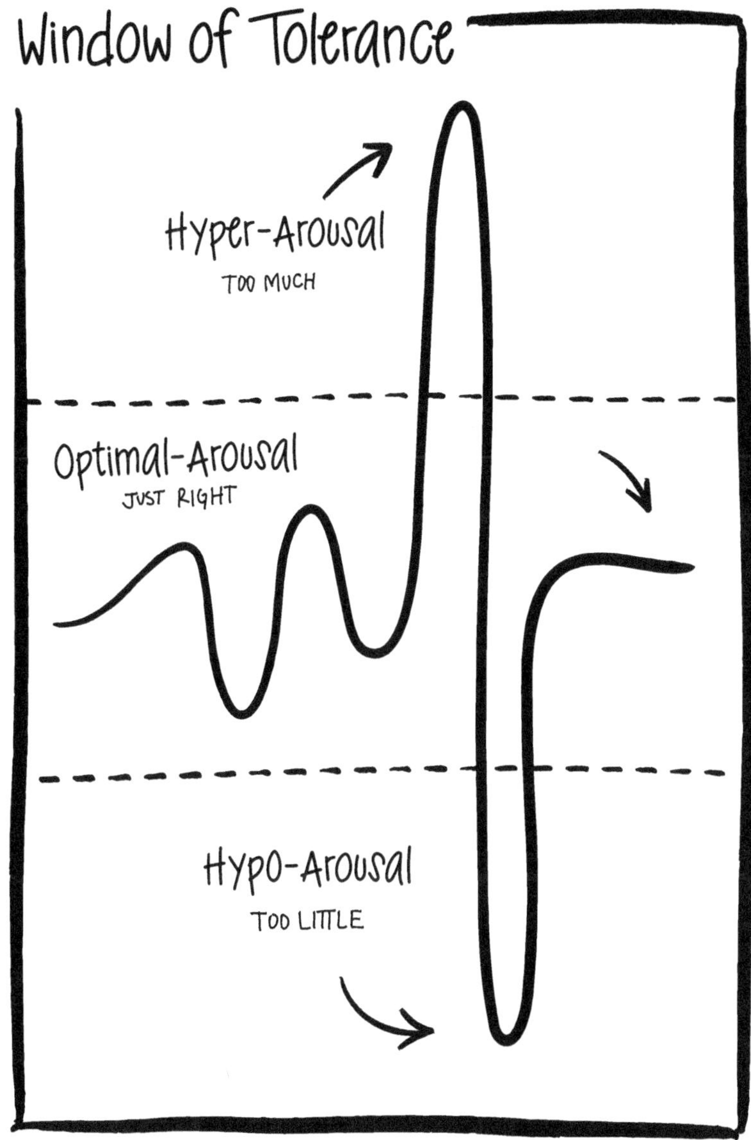

Print It, Post It, Pass It On: This is a great page to stick on your fridge as a go-to resource. Use it to ask your kid or teen where they are at and what they need.

Quick Tip(s): Watch for signs that your child is approaching the edge of their **window**:

- Knuckle cracking
- Fidgeting
- Staring off into space
- Provoking other children
- Whining, etc., all the way up through biting, spitting, and kicking

These are signals, not just bad behavior.

Brain Bite: Young kids find it really hard to notice when they are reaching the edge of their **window of tolerance**, so until they can, it's up to us to notice the signs. When we do, we can intervene quickly before our kids completely exit their **window** and it becomes much harder to regulate or calm them. This is called co-regulation, and it's one of the most important things we can do for our kids.

Co-Regulation with infants and toddlers looks like:

- Rocking, holding, soothing, humming
- Dancing or jumping with them

Co-Regulation with young children and kids looks like:

- Hugging, sitting with, taking deep breaths together
- Dancing or jumping with them
- Taking a walk together

Co-Regulation with older kids or teens looks like:

- Hugging, gentle touch on shoulder, sitting close by
- Taking deep breaths together
- Dancing or listening to music together
- Taking a walk together

DIGITAL DISRUPTION!

Kids and teens get hyper-focused when there is digital media in front of them, be it a "life or death" gaming moment, an argument unfolding on social media, their favorite YouTube video or even waiting their turn. This means that they aren't paying attention to their window of tolerance.

Real Talk

Malik is 8. Here, his uncle describes a recent situation that he experienced with him:

"Malik was waiting to have a turn on the tablet, and I noticed him starting to get red-faced and his fists balling up. Two minutes later he stormed off and as he did so he slammed into his younger cousin who was currently playing on the tablet."

Here's what to say to a young child to help develop this skill (these should be said using a playful, light tone of voice):

"Oh wow, your face is all scrunched up and red! Let's get a drink of water to help us calm down."

"Your body is getting all wiggly sitting down! Let's stand up and shake our wiggles out! Then we can try and sit quietly again."

Or in Malik's case:

"Hey Malik, phew, your face is getting all red! It's hard to wait, I get it. Let's get a drink of water and a snack while your cousin finishes her turn."

Older children and adolescents are much more capable of recognizing initial signs of dysregulation; however, there will still be moments when they struggle with this (remember our earlier example, even adults become dysregulated sometimes!)

When this happens, it is important that the adult maintains the same level of support as they would for a younger child, but the things they say might look and sound different.

Here's what to say to an older child or adolescent to help them recognize when they are reaching the edge of their **window of tolerance** (all of these should be said in a calm, friendly voice):

"It can help our bodies to relax when we take a drink after playing hard."

"I notice you talk less when there are lots of people in the room. It seems like taking a walk helps you feel more relaxed; why don't we do that now?"

"A few times now I've seen you crack your neck and pace around the room before you start throwing things; have you noticed that too?"

Don't worry, eventually children and adolescents will begin to see the signs independently because of these repeated experiences, but until then, don't lecture; offer a sensory activity and save the talk for later.

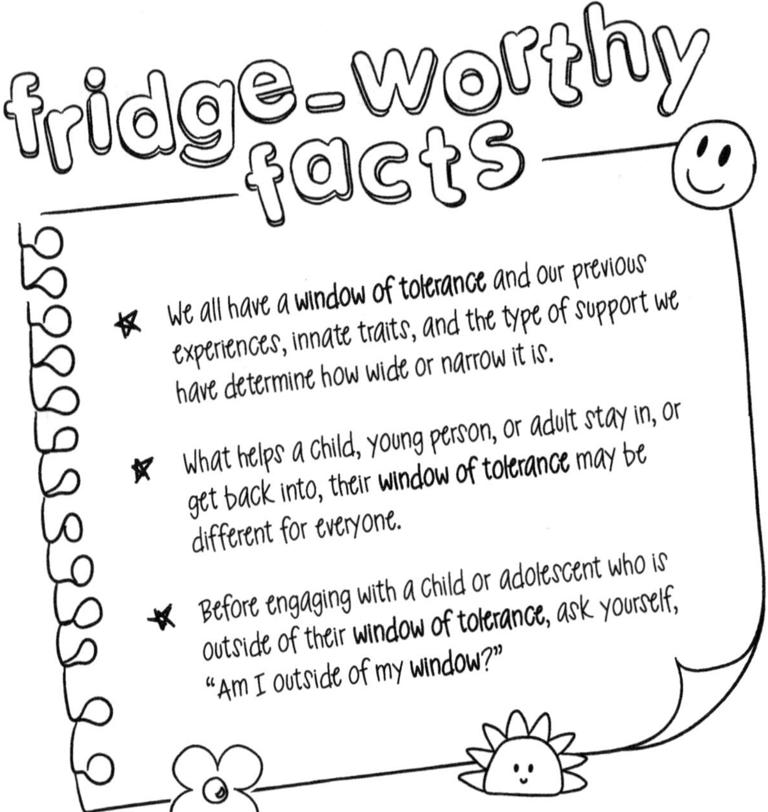
fridge-worthy facts
We all have a window of tolerance and our previous experiences, innate traits, and the type of support we have determine how wide or narrow it is.
What helps a child, young person, or adult stay in, or get back into, their window of tolerance may be different for everyone.
Before engaging with a child or adolescent who is outside of their window of tolerance, ask yourself, "Am I outside of my window?"

CONCEPT 3
Structure and Nurture

"Predictable limits wrapped in a relationship"

Early life milestones that show up here: Attachment, Psychological Home Base, and Power vs. Powerlessness.

Brain Bite: A balance of **structure** and **nurture** is useful in both Upstairs and Downstairs Brain (it helps us reinforce expectations in a way that's easier to digest), but it's especially helpful when your child or adolescent is in their Downstairs Brain and they need someone to help them feel safe *and* loved.

Do you remember your teachers at school? Specifically, do you remember those teachers who were so laid back, chatted to you like a buddy, hung out while you did your schoolwork, maybe let you get away with talking a bit too much? Or how about the teachers who were more like military drill sergeants, who insisted that the rules be followed precisely and there was no smiling allowed?

How did you feel being in each of their classes?

These are the extreme versions of **structure** and **nurture**—terms that we often refer to when working with children and young people.

Now, think about the teacher who you could smile with, who made you feel cared for but who also maintained the rules and expectations so you always knew where you stood with them. How did it feel to be in their classroom?

Structure refers to the limits, boundaries, and expectations an adult consistently provides a child—this helps a child feel safe (emotionally and physically) and trust that the adults can meet their needs.

Nurture refers to being **attuned** to a child and providing care accordingly—it also softens the edge of the **structure** being set.

Too much **structure**? Kids feel controlled.

Too much **nurture**? Kids don't feel safe.

The sweet spot? Predictable limits wrapped in a relationship.

It's important to find a balance of both **structure** and **nurture**. If you tend to use more **structure**, that's okay; just increase the **nurture** you provide, and if you lean toward **nurture** a little too much, just increase your levels of **structure**.

Structure

Structure and expectations are essential. Both will help your child or teen feel safe and secure.

Helpful ways to increase (and maintain) **structure**:

- Keep it simple.
- Set realistic rules.
- Be consistent.
- Follow through.
- Remind often.

"Okay, so how do I say it?"

Oftentimes, the way we set an expectation or limit with a child can make or break the child's ability to hear and follow through with what we're asking.

Remember:

Force + Power + Unpredictable = Defiance and Escalation

Calm + Firm + Consistent = Safety and Security

HELPFUL STRUCTURE FOR ELECTRONICS:

Simple: “Your laptop can only be used for schoolwork from now on.”

Realistic: “I will be checking the laptop every evening before bed to make sure it’s only being used for schoolwork.”

Consistent: “I know you would like to be able to use that website, but because you have been misusing your laptop, that’s not possible.”

Follow through: “I’m so glad you only used your laptop for schoolwork today. I will still be checking it every evening, like I said I would.”

Remind often: “Hey, don’t forget the laptop is just for schoolwork.”

UNHELPFUL STRUCTURE FOR ELECTRONICS:

Complicated: "You can only have the laptop when I say so, but on the weekend you can have it for an hour. You can only watch videos for twenty minutes. You can't have access to social media, but you can look at educational websites."

Unrealistic: "You're never getting your laptop back again."

Inconsistent: "Just for today you can have the laptop for as long as you want."

No follow-through: "Yeah, you can have your laptop."

No reminders: "I told you last week you couldn't play video games after school, but every day you've played video games!"

Below are some examples of what to say to reinforce **structure** and limits:

Child: Give me the tablet!

Parent: I know it's hard to stop doing something you are really enjoying. It's dinnertime and our rule is no screens at the table. Would you like to tell jokes or stories while we eat dinner? Or you can eat quietly. What is your choice?

Child: I'm not getting off the tablet!

Parent: I know it's hard to stop something you are really enjoying. Our rule was twenty minutes of tablet time a day. You can choose to stop now and get tablet time tomorrow, or you can choose to lose your tablet time tomorrow. What do you choose?

Brain Bite: Your kids are going to challenge you and push you to your limits, and when they do, it can be really hard to stay calm (trust us, we've been there). There's a scientific reason for that. When we sense a loss of control, we retreat into our Downstairs Brain (just like our kids), and our words and actions become illogical and unreasonable.

This is when following through is vital. But don't forget, it's okay to say, "Phew, I think we all need a few minutes to calm down," in order to do so.

And try not to be discouraged when your child challenges, or flat out refuses the **structure** and limits you set the first time . . . the third time . . . the tenth time . . . or even the fifteenth time! These skills take time, repetition, and consistency to build. It is normal to see scattered progress.

We promise setting, and accepting, **structure** gets easier the more times you do it!

Nurture

In healthy relationships, **structure** cannot exist without a strong balance of **nurture**. A child must know they are important, loved, valued, and cared for by their caregivers to accept **structure**. We find it helpful to remember, "connection before correction."

What was that Mary Poppins used to say? *"Just a spoonful of sugar helps the medicine go down…"* **Nurture** is the sugar to our **structure** medicine.

Brain Bite: **Nurture** allows the child to separate "I am bad, unlovable, unchangeable" (a shame response) from "my behavior is bad and needs to change" (a guilt response).

(Hughes 2006; Hughes and Baylin 2012).

GUILT VS. SHAME

When we feel shame, we feel bad about ourselves.

When we feel guilt, we feel bad about something we have done, but not about who we are as a person.

Shame is painful, unhelpful, and not productive.

Guilt is also painful, but it helps us to adapt our behaviors to something better.

In a world that seems to have gone from one extreme (authoritarian parenting) to the other (permissive parenting), we are here to remind you that you *can* love your child (provide **nurture**) *and* not be okay with their behavior choices (provide **structure**).

This is exactly the approach we recommend.

Real Talk

This is Sam, she's an elementary school teacher. Here, she is describing how she balances **structure** and **nurture** with her first graders:

*"The kids are expected to walk quietly down the hallway in a line, one behind the other, without touching the walls. Well, for some kiddos those rules (**structure**) are manageable, but for others it's quite tricky and they become wiggly, loud, or begin touching everything. Instead of yelling out expectations ('Hey! Quit touching everything! Keep your hands to yourself!') from across the hallway, we began walking next to, or even holding hands with, those kids*

who found the rules hard to follow. And we noticed that their bodies started to relax and they could follow the rules better."

The close proximity or gentle touch that Sam is talking about conveys this message to her kids, *"I see that this is hard for you right now, but don't worry, I'm here to help you because I care about you."*

It's important to know that providing **nurture** doesn't mean treating the child or adolescent like a baby. Nor does it mean allowing them to do whatever they want. **Nurture** will look different for every kid (and every adult), and it doesn't always have to involve touch, sometimes a simple "I see you" or "I care about you" is enough.

Helpful Ways to Increase (and Maintain) Nurture in a Digital World:

- High fives and fist bumps when your kids or teens follow through with expectations
- Snuggling or sitting close by *before* providing reminders to your child that it's time to turn off the tablet
- Sitting down next to your teen when you talk about expectations and boundaries surrounding social media use
- Holding hands when you are trying to encourage your younger child away from digital media
- Using a transition activity—sing a song, code word or phrase, or a fun cooperative game when moving away from technology

- Using a gentle tone and kind body language to convey empathy when explaining to your child why they can't play certain video games

We understand that being nurturing can sometimes feel like we are "rewarding" a negative behavior, but we promise you that is not the case. It's simply a way to provide support to our kids as they learn to understand and remember our expectations.

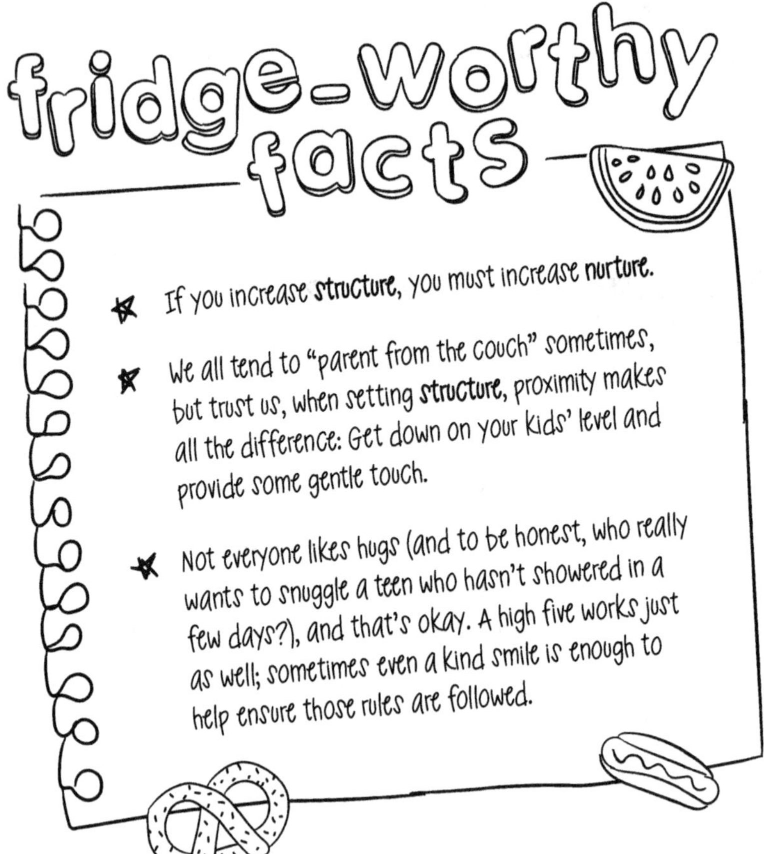

CONCEPT 4
The Importance of Play

"Play is the work of the child." —Maria Montessori

Early life milestones that show up here: Regulation of State, Mastery, Power vs. Powerlessness.

Brain Bite: **Play** is especially helpful when your child or adolescent is in their Midbrain and they need help staying out of their Downstairs Brain while also trying to access the neat functions of their Upstairs Brain.

Think about the last time you had a really long, busy day at work. What did you do afterward to unwind? Maybe you took a walk or went to the gym. Maybe you met a friend for dinner or to watch a movie. Maybe you put some music on and danced. The proverb "all work and no play makes Jack a dull boy" is true for almost all of us: We feel much better when there's "play" involved.

Now, think about the last time you had fun while learning something new. How long did it take for your skills to improve? What about the last time you had to learn something really tedious, which didn't involve any fun. How long did it take for you to master?

Our experience has taught us that **play** can serve many purposes: It can teach skills, build self-esteem, and create strong relationships between adults and children. We have also found that when **play** is involved, kids learn faster—and we're not just saying that; it's science!

Brain Bite: The late (and wonderful) Dr. Karyn Purvis apparently stated that it typically takes around four hundred repetitions to create a new connection in the brain, but when learning occurs through **play** it only takes about ten to twenty repetitions to achieve the same result; essentially meaning that **play** significantly speeds up the process of building brain connections. While the research to support this might be missing, our personal experiences, and those with the kids and families we treat, tell us that this may in fact be true.

And there's more!

Brain Bite: When we laugh and engage joyously "in the moment" with another person, it's hard to stay in our Downstairs Brain. Our survival responses can rest, meaning our brains are better able to take in new or different experiences (Isenberg & Quisenberry, 2002).

A certain amount of independent play is healthy, but unfortunately you can't just put some toys in front of your child and expect them to learn something new or do something differently (if only that were the case, right? We are all parents here; we get it). This is especially true when it comes to managing digital media and digital disruption.

Instead, we need you to be **playing** *with* your child or teen.

💡 **Quick Tip(s):** Instead of giving your 6-year-old a tablet with an educational app on it and leaving them alone, use the app together, talk to your child about what they are learning, quiz them, and pretend you don't know the answers but they do. Or pull up a dance video on the internet and challenge them to a dance-off (particularly helpful if they have been sitting all day!).

💡 **Quick Tip(s):** Your teenager is learning about a subject at school and they have been reading from their textbook. You notice their shoulders are slumped and their face looks grumpy.

Look up a fun video on the subject they are learning and watch it together, asking questions afterward to check for understanding.

Play is a great way to use some of the benefits of technology, rather than trying to exclude it altogether. Similarly, we can also use **play** to help diffuse some of the challenging behaviors we encounter with digital media.

Real Talk

Riley is 7. She finds it hard to transition from screen time to another activity. Instead of just removing the screen from Riley, her dad shares that he wants to try something different, something fun:

"I knew I could just remove the screen and we could risk a meltdown, so instead I asked Riley what her game was about. She told me it was a racing game. I asked her if we could each race and see who got the fastest time, with the winner getting two marshmallows as their prize. She was all about it, saying she was going to beat me. I told her, 'Okay, and then after it's time to do our chores.' She beat me of course, got her marshmallows, and did her chores! Meltdown avoided!"

Real Talk

Etta is 16. She spends a lot of time making video calls to her friends and is always putting on a lot of makeup and taking selfies. Etta's mom worries that her daughter's self-esteem is low and doesn't want her spending so much time on her phone. Here, she describes the playful approach she used to talk to Etta about this, rather than just taking her phone away:

"With Etta being 16, I knew that her phone was important to her and the way she connected with friends, but I was concerned about how much she was on it and how much makeup she was applying to 'look good.' Instead of just telling her to quit, I asked her to give me a makeup tutorial! She was so excited to make-over her 'old' mom! While she did it, I talked to her about my concerns and she really opened up to me. She even agreed to limit her call time and ease up on the makeup."

Don't forget, just because we are having fun and creating joyful experiences with our kids it doesn't mean that we accept hurtful or unhealthy behaviors. Instead, we understand the way that **play** can make it easier (and sometimes quicker) for expectations and limits to be learned and followed-through with. (You'll find plenty more examples of how to use **play** in our scenarios in part 2.)

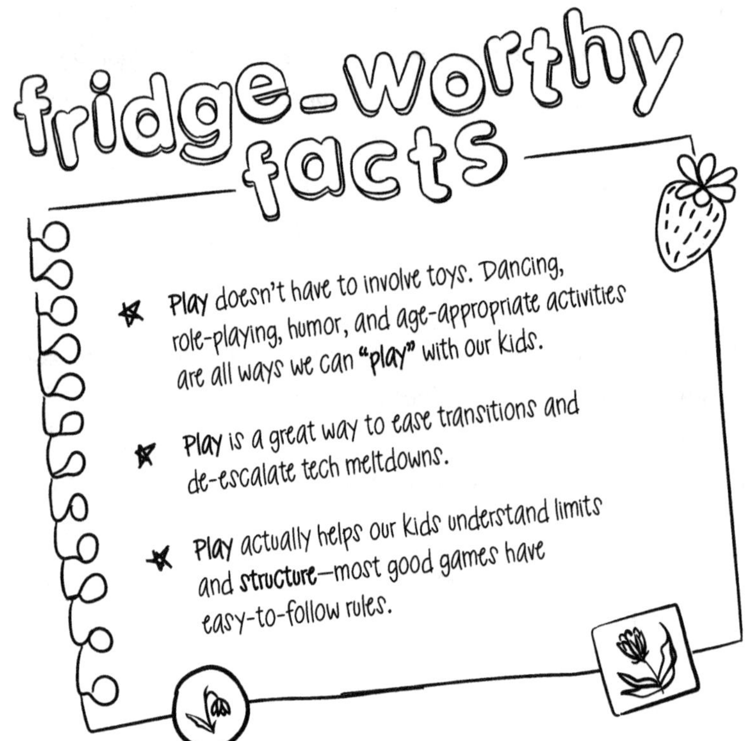
fridge-worthy facts
★ **Play** doesn't have to involve toys. Dancing, role-playing, humor, and age-appropriate activities are all ways we can "**play**" with our kids.
★ **Play** is a great way to ease transitions and de-escalate tech meltdowns.
★ **Play** actually helps our kids understand limits and **structure**—most good games have easy-to-follow rules.

CONCEPT 5
Verbal Responses

"Validate first, advise later"

Early life milestones that show up here: Attachment, Psychological Home Base, Regulation of State, Object Constancy, and Power vs. Powerlessness.

Brain Bite: Use of **verbal responses** is especially helpful when your child or adolescent is in their Midbrain and has a lot of big feelings.

Have you ever experienced someone telling you "Oh, it's not that scary!" or "You just need to get over it," during a time when you felt very anxious or stressed about something? How did they make you feel? Would it have felt different for you if they had instead said, "I heard you say that this is a really worrying time for you. I understand where you are coming from."?

If, like me, you're a "fixer," that is, someone who just really wants to make a person feel better by helping to solve the problem, then it's likely that at some point, you have dismissed someone's feelings. Don't feel bad; it's not done intentionally—we just don't know what we don't know. I, for one, am definitely guilty of this.

Check out the video "It's not about the nail" for a humorous take on "fixers."

The trouble is, if a person doesn't feel heard and understood, often they will continue to express their feelings (we call this "venting"), until they do. Essentially, they get "stuck" in venting, without ever being able to move toward a solution. By the way, fixers also vent from time to time too.

In case they haven't made it clear to you already, our kids and teens *really* need to feel heard and understood. Our reward for doing so? Having a child who trusts us and feels emotionally safe with us. If only it was that simple, right?

What we are looking for doesn't require superpowers, but it does take some practice. There's skill involved in giving **verbal responses** that allow for venting (without getting stuck there), validation that's genuine, and identification of solutions, but I promise you don't need a magic wand.

When giving **verbal responses**, our goal is to put words to what our kids are experiencing in their bodies, what they're communicating through their behavior, and help them to make sense of their reactions when they have big feelings.

But wait, I know what most of you are thinking, *My kid just says "I don't know" when I ask them how they are feeling!*

Brain Bite: It's normal for a child or adolescent to respond with "I don't know" when asked about how they are feeling.

Children and adolescents often have difficulty in identifying and expressing their emotions. They frequently say they "don't know" what they're feeling, state they are "mad" for every feeling they experience, struggle to identify emotions in the correct context, or state they "don't have feelings."

Sometimes kids also avoid talking about emotions because they feel vulnerable, but at other times the child genuinely does not know what they are feeling and is embarrassed or shy to admit they do not have an answer. Haven't we all felt like that before?

Real Talk

George is 13. Here's George's dad describing a recent interaction with George:

"I noticed that when George would get mad about something, he wasn't able to tell me; instead, he would physically lash out by hitting things. When I asked him why he did that, he said, 'That's what they do in the movies I watch.'"

In this particular example, we are dealing with two challenges: (1) George's developmental ability to verbalize emotions, and (2) the unhelpful skills he has learned through screen time.

The key to **verbal responses** is this: We are connecting body sensations to emotional language while also assuring the child that their emotions (*not* their behavior) make sense in a validating, accepting, and empathetic manner.

Let's consider the above example and how we might respond. To help you, we have identified three types of **verbal responses** you can use in this situation: **Honoring**, **Wondering**, and **Validating**.

Dad: I heard you say that sometimes you feel really mad, is that right?

George: Yes.

"And it seems like when you get mad, you hit things, like you've seen them do in the movies, is that right?"

"Yes."

"I wonder, do you think it's okay that they hit things in the movies?"

"Well, they don't get into trouble for it."

"You're right, sometimes they don't get into trouble. I can see why you might react that way when you feel mad now."

(This is called **Honoring** their experience and helps us convey to our child that we get why they might think or act a certain way—we are not saying it's the right way, we are just saying that we understand why they might think it's okay to act a certain way.)

"But movies aren't real life, are they?"

"No."

"What usually happens in real life when we hit things?"

"We get into trouble."

"Yeah. I wonder if you really feel so mad that you want to hit things, or if you would just like someone to help you feel less mad?"

(This is called **Wondering**, and it helps kids feel accepted and reduces defensiveness.)

"Yeah, but I feel so mad sometimes that words don't come out."

"That's okay, I think I've felt like that too sometimes. I want to yell the words out but they get stuck and I find that my fists ball up like this."

"You do?"

"Yeah, I think it's because the thinking part of my brain turns off in those moments, so it's hard to get the words out."

"Oh."

"Yeah, so it makes sense that you can't get your words out."

(This is called **Validating** their perspective, and it helps create a calming sense of security to someone who feels misunderstood, invalidated, or otherwise compromised in a conflict.)

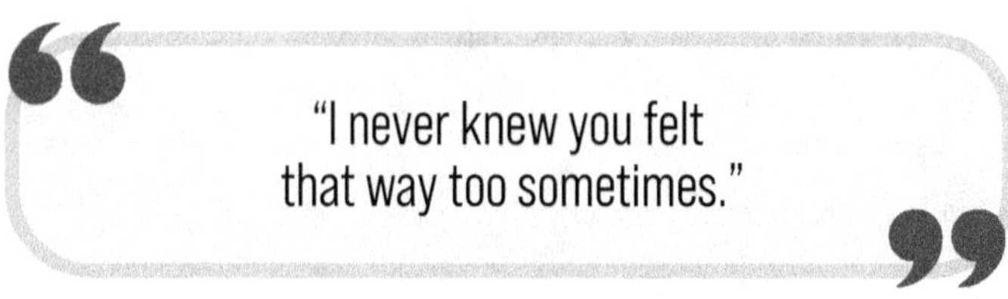

Honoring: "That behavior made sense to you from your perspective."

Wondering: "I wonder if you're mad or scared?"

Validating: "That makes sense; I've felt that way too."

Our goal with our **verbal responses** is to de-escalate, or calm, the situation by relating to our child and conveying our understanding, not providing solutions—our kids aren't ready for that yet. We want to make sure that they are out of their Downstairs Brain before we try to address the conflict; otherwise, they may struggle to remember what we talked about.

When we are sure that our child is ready, here is how we might find a solution while maintaining our expectations:

"You know, I'm here to help you when you feel like that, if you would like?"

"Yes."

"Okay, so when you're mad, I've noticed that your face starts to get a bit red and you pace around more."

"I do?"

"Yes. I wonder if it would be helpful for me to give you a signal when I notice that happening?"

"Okay."

"Maybe I could say, 'Oh there's some mad words trying to come out!' Would that be helpful?"

"Haha yes, okay."

"It's okay that you feel mad sometimes, but it's not okay to hit when you do. Can you understand that?"

"Yes."

"And it's important to remember that not everything you see on the TV or in the movies is the right way to do things.'

"Yeah, I know, that was kinda silly."

"Okay, I'm glad we had this talk. I'll try my best to help you next time I spot your body trying to tell you it's feeling mad."

AN IMPORTANT NOTE ABOUT ADULT REGULATION

If you are feeling angry or overwhelmed during an interaction with your child or teen, the chances are you're on your way downstairs, in your brain that is. If you have listened to your child tantrum all day it can be very difficult to genuinely validate their perspective while putting your own emotions to the side. Likewise, when you have put in diligent effort with your teen only to continually be met with resistance, anger, and (dare I say) attitude-fueled responses, it can feel exhausting, and sometimes impossible, to continue to show up with validation and empathy for what your child might be feeling and remain **attuned** to their needs.

Trust us, as parents and caregivers, we get it. That's why it's really important to be self-aware and know what sets you off and what helps you stay regulated.

Here are some recommendations for keeping your cool, and remaining "upstairs":

Know Your Buttons: Does bickering make your blood boil? Do you see red when your kid is refusing? Do you cry when there's cursing? All jokes aside, knowing what behaviors or actions push your buttons is the first step in staying calm.

Know What Works for What Button: When you're in the middle of a tech tantrum it might be hard to use your go-to stress relief tactics such as taking a walk or listening to music, so it's important to find a quick, in-the-moment tool to help you keep calm. Personally, I like to take deep breaths and count to five.

Know When to Tap Out: Sometimes our emotions just get the better of us, and there's no shame in that. We've all been there (and will be again). Tapping out doesn't mean you are "admitting defeat," nor by doing so are you saying your child's behavior is okay. It might even be an opportunity to model stepping away from a challenging situation in a healthy, respectful way.

Quick Tip(s): You can "tap out" by asking for help or support from another adult, taking a break for a minute to get a drink, or sitting down and taking a few deep breaths.

Here's what you can say if you recognize you need a break:

"You know what, I am noticing I am really thirsty. I need a drink of water. I'm going to get one. Do you want me to bring you a drink too?"

"Hey, [another adult in the area], can you sit with [child] for a minute while I get us a snack? I think we could both use a snack right now."

"I feel frustrated right now and I'm going to take a break. I'm going to go downstairs and load the dishwasher. I'll be back to check on you in ten minutes."

SOME ADDITIONAL, ADULTS-ONLY POINTERS

- Occasionally our kids need to tap out from us too. Sometimes, when a child is angry with us and we are their "target," it can be difficult for them to engage in any kind of regulating activity with us. Having another adult step in can help you and your child calm down.
- Be open to someone else tapping you out. Phew, this is a hard one for us grown-ups, but sometimes, we're just not going to notice when we've "hit our limit," are power

struggling, arguing, or becoming too emotional in an interaction with the kids. Just know, it happens to the best of us and try not to take it personally when someone steps in to offer help.

Always remember, if you tap out, you must tap back in and reconnect at a time when you are ready; this helps our relationships with our kids stay strong.

CONCEPT 6
Sitting in the Yuck

"Being comfortable with being uncomfortable"

Early life milestones that show up here: Attachment, Psychological Home Base, Regulation of State, Object Constancy, Power vs. Powerlessness.

Brain Bite: **Sitting in the yuck** is especially helpful when your child or adolescent is in their Midbrain and is experiencing feelings or situations that are making them uncomfortable.

Remember that one time, probably sometime in your teens, when you made a pretty big mistake? Like driving into the back of another car when you were singing along to the radio with your friends instead of paying attention? Or when you stayed out past your curfew but couldn't get back in the house without waking your parents to unlock the door? Perhaps you even snuck out to a party but ended up needing someone to pick you up so you had to call home? I think we can all recall that sinking feeling associated with such mistakes, the pit that forms in your stomach and the dread that comes from not knowing what's going to happen next. It's not pleasant; you might even say it's "yucky."

None of us *want* to feel like this, but sometimes life just doesn't work out that way. By adulthood, most of us become better at tolerating feelings that don't feel good, but kids have to build their **window of tolerance** for safely sitting in big or overwhelming emotions so they avoid unhelpful responses such as anger or shutting down. As parents, we have to be prepared to support our kids through these often-uncomfortable Midbrain emotional experiences.

Next-Level Knowledge: Take a look back at page 91 if you need a reminder about your kid's **window of tolerance**.

We know it can be hard to see your child or teen feel deeply sad or disappointed and sometimes that makes us want to justify, rationalize or "fix it" for them. So just to be clear, "**sitting in the yuck**" does not mean we are resolving, or "making it better." Instead, think of this as simply *being with* and patiently supporting, while your child works through their difficult emotions.

Brain Bite: When a child or adolescent doesn't quite have access to their Upstairs Brain, they find it hard to **problem-solve** and reason. Instead, they are fully absorbed in simply expressing their emotions.

Real Talk

Here's Jada's auntie, telling us about a recent interaction with Jada, who's 15:

"Jada was telling me that someone she thought was a friend had said something unkind about her on social media and lots of people had seen it. I told her that she just shouldn't be friends with them anymore and that she should just ignore what is said on social media. It seemed like Jada didn't even hear me as she kept asking me why her friend would do that and that she just didn't understand. I wasn't sure what to say; I just wanted to fix the problem for her."

Jada's auntie had the best intention—she wanted to make her niece feel better. But in the moment, Jada needed her auntie to just listen and sit with her while she expressed the feelings she was having. Here's what Jada had to say about the interaction:

"I know my auntie was trying to help, but it felt like she wasn't hearing me. Deep down I knew that I probably wasn't going to be friends with that person anymore, but it really upset me and I was worried that I couldn't use social media anymore and see all my other friends."

Sometimes "**sitting in the yuck**" also means helping our kids understand the **natural and logical consequences** of behavior choices; this is often the case when it comes to screens, technology, and digital media use.

Let's consider another example:

Lucas is 12. He loves playing games on his tablet. In his house, there are simple rules when it comes to tablet use that Lucas knows and understands and can usually follow. One evening, however, Lucas was having a hard time turning the tablet off when it was time to join his family for dinner:

Dad: Hey, kiddo, in five minutes dinner will be ready and it will be time to turn off the tablet.

Lucas: Uh-huh

"Okay, Lucas, remember our rule, you get two minutes and then the tablet is switched off; otherwise, no tablet time tomorrow."

"I just need to complete this level!"

"I'm starting our timer now."

Timer goes off.

"Okay, switch it off and sit at the table."

"No! I need to finish this level!"

"You can put the tablet up or I can put it up, what do you choose?"

"I'm nearly finished!"

"Okay, I will put the tablet up for you."

"No, Dad! I will do it!"

Lucas continues to play while taking his tablet to the area where they store them. Eventually, Lucas turns the tablet off and joins his family at the dinner table.

"Lucas, our rule is that you have two minutes and then the tablet gets switched off. Did you follow that rule?"

"I tried but I just needed to finish that level!"

"I'm glad that you tried but you didn't follow our rule."

"I want my tablet tomorrow!"

"I hear you; it will be hard not to have it."

(Lucas is crying.)

"I know it makes you sad to not be able to have your tablet."

"Please can I have it?"

"I know that your games are fun and important to you. We also have to make sure we are sticking to our family rules, so tomorrow we will find a different activity to do instead of our tablet."

Next-Level Knowledge: You'll notice that riding out emotions such as disappointment, sadness, and anger often includes using a combination of the **verbal responses** discussed in the last concept: validating their perspective, honoring their experience,

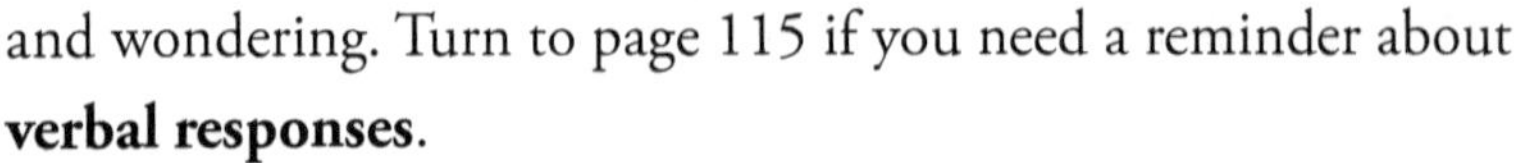

and wondering. Turn to page 115 if you need a reminder about **verbal responses**.

Now, I can already hear you saying, *"That's easier said than done!"* And you're right. It takes a lot of deep breaths, patience, and consistency to make it through a tween tech tantrum. It also takes trial and error, and follow-through. This won't be the last time Lucas's dad will have to stick with, and reinforce, the rules, while sitting with his son through his sadness and frustration.

Quick Tip(s): Here are a couple more go-to things to say while **sitting in the yuck** with your child:

"I know this wasn't the answer you wanted. I'm here to work through it when you're ready."

"I can only imagine you're feeling sad and disappointed to miss the sleepover this weekend. I get how you're feeling, and I'm ready to talk when you are."

"I know this is hard. It is understandable that you are upset when you lose privileges."

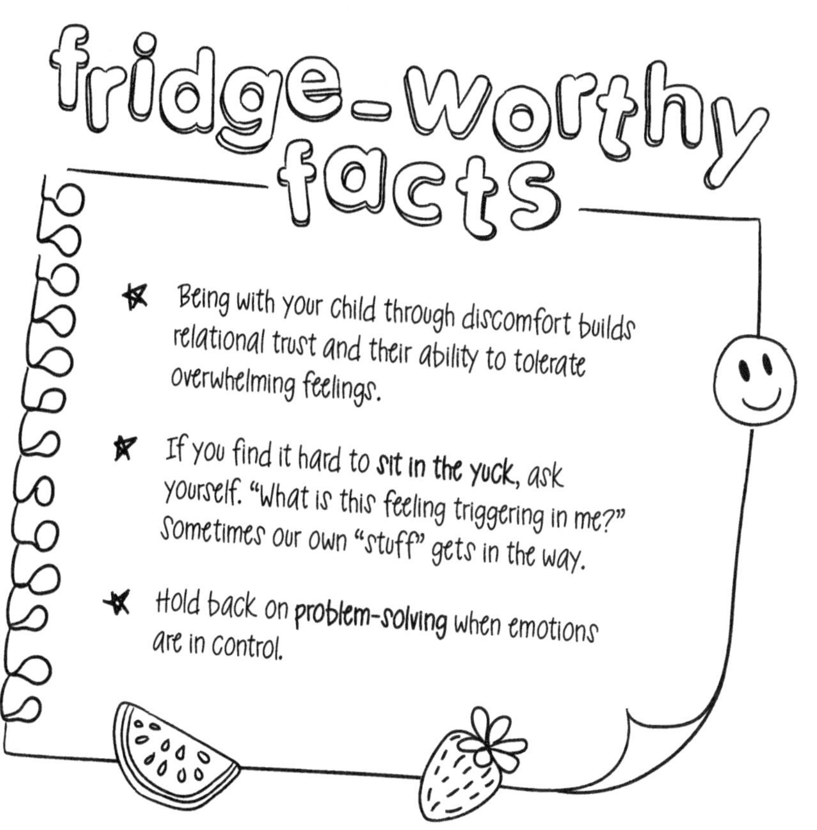
fridge-worthy facts
★ Being with your child through discomfort builds relational trust and their ability to tolerate overwhelming feelings.
★ If you find it hard to **sit in the yuck**, ask yourself. "What is this feeling triggering in me?" Sometimes our own "stuff" gets in the way.
★ Hold back on **problem-solving** when emotions are in control.

CONCEPT 7
Problem-Solving and Abstract Thinking

"Learning to think beyond the moment"

Early life milestones that show up here:
Attachment, Psychological Home Base, Object Constancy, Mastery, Power vs. Powerlessness

---

Brain Bite: **Problem-solving** and **abstract thinking** is especially helpful when your child or adolescent is in their Upstairs Brain and needs help moving from feelings and behaviors to solutions.

Have you ever been really frustrated with a work colleague because they didn't do or finish a task properly? After calming down you had time to think clearly and gather more information. You found out they were dealing with another stressor at work that you didn't know about, which impacted their ability to do the task you asked of them. With this new understanding you were able to communicate with your colleague in a calm, clear way that expressed your feelings but also helped resolve the situation in a kind and empathic way.

Problem-solving and **abstract thinking** is all about taking what you already know and applying it to new or different situations in a way that is informed, regulated, and helpful. For our kids, this often means understanding how thoughts and feelings impact behaviors and using that to have healthy, positive, and productive relationships and interactions in life.

Real Talk

Remember 13-year-old George, who struggled to say when he was feeling mad and instead would physically lash out and hit things (if you need a reminder you'll find George on page 117)? Using **verbal responses**, George's dad helped George connect his feelings and behaviors together. A few weeks after we spoke to them, George's dad called us to tell us about something neat that happened:

"I noticed George starting to get frustrated with a video game he was playing. I was about to give him the signal that his body was getting mad when he jumped up and said, "Ugh I'm so mad! I'm going outside to cool off." I couldn't believe it; inside I was doing a happy dance. But being the cool dad I was, I just said, "Okay, great idea!" and let him go."

Now, if you've ever found yourself in such a situation you probably had the exact same thought as us: *"Wait, what just happened? Did I hear them right? Who are you and what did you do with my teen?!"* When we spend so much time helping our kids connect their feelings and behaviors it can be a bit startling when they finally start doing it themselves!

This is a moment to celebrate!

Brain Bite: A consistent approach will strengthen the staircase between your child's Downstairs Brain and Upstairs Brain, allowing them to develop the ability to recognize and better express what they are thinking and feeling. Now they can start using those Upstairs Brain skills! When the Upstairs Brain is activated, you will notice that your child is much better able to communicate clearly, think outside of the box, and **problem-solve** with more independence. **Abstract thinking** skills also develop and include delayed gratification, future thinking (applying logic to events that haven't happened yet), taking accountability for their actions, and understanding other's perspectives.

Brain Bite: These competencies are more advanced versions of the early milestones of Object Constancy and Power vs.

Powerlessness (look back at chapter X if you need a reminder about these).

Let's look at what George's dad could do next to help George work on his ability to think about the future:

"Hey, George, I am so glad you were able to recognize how mad you were and take a break. I wonder what might have happened if you hadn't?"

"I would have thrown that stupid thing against the wall!"

"Oh wow, yeah that probably would have broken it and then you would have been really sad I bet."

"Yeah, I'm glad I walked away."

"Me too, good job."

A NOTE ABOUT HOW KIDS SAY SORRY

While our kids' Upstairs Brain is still developing, it's important to remember that it can be hard for them to take accountability for their actions, for a number of reasons, including embarrassment, anger, or anxiety about consequences. And they definitely won't stand up and give you a thirty-minute presentation on all the ways they are sorry.

As adults, we have to learn to accept that apologies, and taking accountability, in kids, and especially teens, may look and sound different to what we would expect from an adult.

Here are some things to listen out for when waiting for an apology:

"I'm sorry."

"I know I shouldn't have done that."

"I could have handled that better."

"I'm so mad at myself."

"I know, I get it."

If you hear these, or other similar words, know that your kiddo is doing their best to make things right, even if it's not exactly how we would do it.

Celebrate these small wins. They're signs of a growing brain.

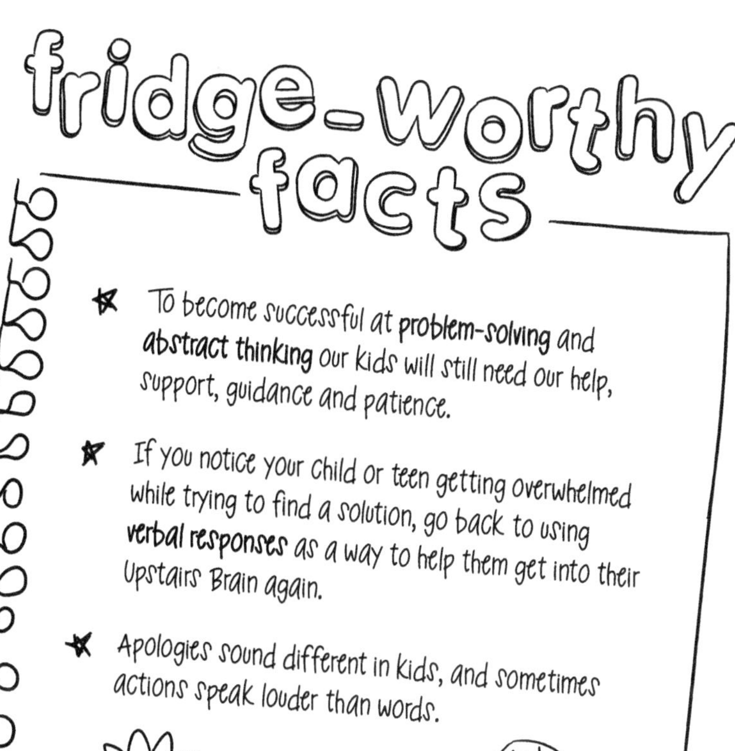
fridge-worthy facts
To become successful at **problem-solving** and **abstract thinking** our kids will still need our help, support, guidance and patience.
If you notice your child or teen getting overwhelmed while trying to find a solution, go back to using **verbal responses** as a way to help them get into their Upstairs Brain again.
Apologies sound different in kids, and sometimes actions speak louder than words.

CONCEPT 8
Independent Regulation

"Finding calm without a coach"

Early life milestones that show up here: Attachment, Psychological Home Base, Regulation of State

---

Brain Bite: **Independent regulation** is especially helpful when your older child or adolescent is in their Upstairs Brain and they don't have a trusted adult around to co-regulate with.

Note: We would not expect a child younger than 5 to be able to **independently regulate**; however, you can encourage them to and see what happens—don't worry if they can't or it only lasts a few seconds!

Remember our earlier example where you were told to sit still for hours and listen to someone speak? Well, imagine now that instead of being told to sit still you were told that you could get up and move or grab a snack or a drink if you needed to.

This time, when the presentation is two hours in and you feel yourself getting restless (that's the edge of your ***window of tolerance****), you get up and pace at the back of the room and grab a glass of water—meeting your physical and movement needs. After a short while you're able to sit back down and focus, without snapping at anyone, for the rest of the presentation.*

Adults with a fully mature brain are typically able to use **independent regulation** skills to get back into their **window of tolerance** where they can listen and learn effectively. However, because our kids and teens brains are still developing, their ability to do the same will be inconsistent and imperfect—and that's okay.

It's important to start small so that your child can build up to managing their brains and bodies independently. And don't forget, digital media and screens can cause distractions that make this harder.

QUICK WAYS TO START SMALL

Instead of this...	Try this...
Giving your child or teen a screen when they have big feelings	Encourage them to take a one-minute walk, dance break, music break, or snack break, then allow them to have the screen. Increase the minutes they have to regulate independently over time.
Leaving your child or teen alone with overwhelming feelings	Engage them in a regulating activity that they enjoy, then ask them to try it by themselves for a few minutes. Afterward, check in with them. Next time, encourage them to try the activity independently.

Real Talk

Grace is 13. Here is her mom telling us about a recent situation that happened at home:

"Grace sits at the kitchen table to do her homework. She was trying to do some hard math problems, and I could hear her getting frustrated. A few minutes later it was quiet so I went to check on her and found her outside on our porch swing. She told me she just needed a few minutes for a break then she would try her homework again. I was so proud of her! Two nights later, Grace was doing homework again. She seemed to be getting frustrated. I was just about to go in and check on her when I heard her throw her textbook on the floor and yell, 'Ugh, this is so stupid!' I couldn't understand why she didn't take a break before getting so mad like she did a few nights ago."

Grace's mom, like most of us in that situation, probably wondered if aliens had come down and changed out her teen overnight, but really, this is just typical "developing-brain" behavior. Much like learning to tie your shoelaces, mastering how to **independently regulate** takes time and practice.

To help Grace, her mom will need to do some minor co-regulating:

"Hey, Grace, oh man, homework again?"

"Yes and it's so stupid!"

"Oh no! Here, let's go sit on the porch and swing for a few minutes."

(Grace's mom is **attuned** to Grace's needs and engages her in a co-regulation activity because she knows that this type of movement helps Grace get back into her **window of tolerance**.)

They swing for a few minutes without talking.

"Let's talk about what happened in the kitchen."

"I know, I threw my book but the teacher went too quickly in class and now I don't understand the homework."

Mom hugs Grace.

(This is a version of **sitting in the yuck**, using **nurture**.)

"Oh honey, I'm sorry to hear that. I know you've been working hard at school and you always get your homework done on time. It surprised me when you threw your book."

"I know, I didn't mean to."

(This is a teenager's apology, just FYI.)

"Thanks for saying that. What could you do next time to stop this from happening again?"

"Well, I should have just told you I was feeling mad."

"Yes, you could do that, but what if I'm not around?"

"Well, being on the swing makes me feel better."

"Yes, I noticed that the other day when you were feeling frustrated about your math homework."

"I could come out here and swing next time to calm down."

"That's a great idea!"

Quick Tip(s): Here are a few more activities that kids and teens can do to **independently regulate**:

- Take a walk
- Go for a bike ride
- Shoot some basketball hoops
- Work on a puzzle
- Listen to some music
- Sit quietly

You could also just ask your child what works for them. If it's an option that's safe, doable, and one that you are okay with, it's a win!

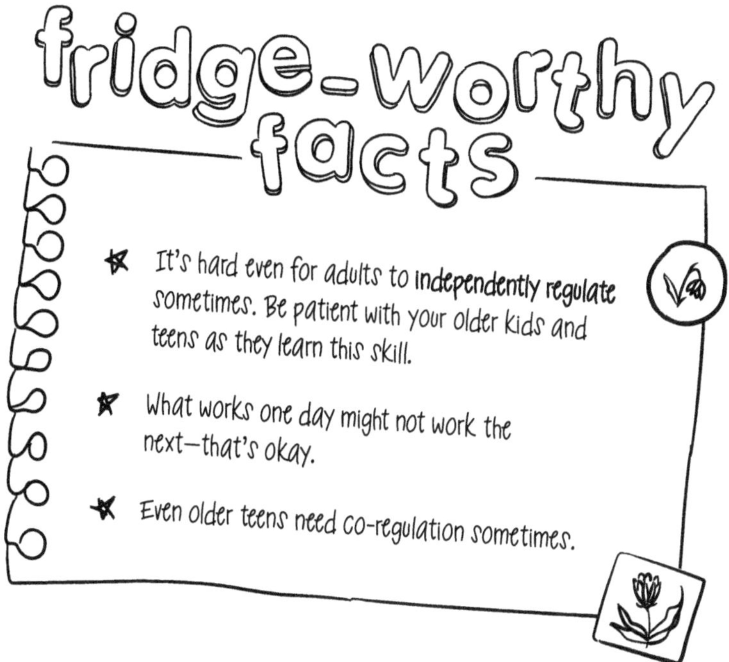

CONCEPT 9
Natural and Logical Consequences

"Teach, don't punish"

Early life milestones that show up here: Attachment, Psychological Home Base, Regulation of State

---

Brain Bite: **Natural and logical consequences** are especially helpful when your child or adolescent is in their Upstairs Brain and they need help understanding that sometimes their words and actions cause negative or unwanted outcomes.

Have you ever left the house without thinking your day through clearly and ended up getting caught in the rain without an umbrella? How about the time you forgot to set your alarm because you fell asleep watching your favorite show, then woke up late for work? Maybe you've even forgotten to take your kids to an event because you didn't write it in the calendar.

So, let's be real, how many of you turned to this section first? It's okay to admit it. We all have moments where we think that if our kids just understood the consequences of their actions then they wouldn't keep behaving in ways that are unhelpful, hurtful, or challenging. Unfortunately, this sometimes means that we throw out unrealistic, and oftentimes unhelpful, punishments instead of **natural and logical consequences** that can help our kids learn and grow.

🧠Brain Bite: When consequences are connected to actions (**natural**) and are both proportional and respectful (**logical**), they effectively establish accountability.

A quick definition will help here:

Natural and logical consequences tend to be their name: They are natural, meaning they would generally and naturally occur in the situation, and they are logical, meaning they make sense in nature, intensity, and duration for the situation or behavior.

Punishments tend to involve something an adult asks a child to do because of how an adult feels about a child's behavior, whether or not it makes sense and/or would naturally occur in the situation.

Real Talk

Jeremy is 9 and repeatedly refuses to turn off his tablet at the end of the allotted time.

Jeremy's dad has two options:

1. **Natural and Logical Consequence** = *"If you choose not to hand in your tablet when it's time, you will not be able to use it tomorrow during your next 'tablet time.'"*

2. Punishment = *"I'm done with this; you'll get it back when I say you get it back."*

We would be lying if we said that we have never uttered Option #2 (or similar)—we all have, and we all will again. When our kids test us, it can be hard to stay out of our Downstairs Brain—we understand. What's important is that you *try your best* to use Option #1 as much as possible. Because while Option #2 feels good for a minute (you are regaining your sense of control over the situation), in the long run it's not realistic; it doesn't make our kids respect us and it doesn't teach them how to better handle having to turn their tablet in.

EXAMPLES OF NATURAL AND LOGICAL CONSEQUENCES VS. PUNISHMENT

Punishment	Natural and Logical Consequence
Scenario: Toddler has a screaming meltdown when screen is turned off.	
Toddler is made to sit in time-out until they are calm.	Caregiver helps toddler co-regulate and recognizes that the toddler needs more help transitioning from screen time to other activities. Screen time is reduced and replaced with reading books or playing with caregiver until this skill is developed more.
Scenario: Child just wants to play video games all day and refuses to do chores by being defiant and angry.	
Child is made to do chores for an entire weekend.	Video games are not allowed until age-appropriate chores have been completed each day/each week. Caregivers provide reminders and check in with child to make sure expectations have been met.
Scenario: Young teen posts something unkind about another person on social media.	
Young teen loses privileges to all social media apps, cell phone, and computers and is forced to write a public apology.	Social media apps and screens are monitored by parents/caregivers for content—you might also consider installing a monitoring app on cell phones and computers. Caregivers engage young teen in conversations about bullying.
Scenario: Older teen is caught looking at websites that contain inappropriate content such as pornography or gambling.	
Older teen is forced to write an essay on why it's wrong to look at inappropriate websites and all technology including gaming consoles are removed indefinitely.	Access to the internet via cell phone, tablet, and computer is restricted, and usage is monitored by caregivers. Caregivers and older teen discuss why accessing inappropriate websites are unhealthy and the issues they might cause and come up with a strategy to manage this behavior.

KEY COMPONENTS OF NATURAL AND LOGICAL CONSEQUENCES:

- Do not determine or give out consequences when you are angry, upset, overwhelmed with, afraid of, or frustrated with a child.
- Regulation and re-connection need to take place before a consequence is given; your child or teen will not be able to understand the consequence if they are in their Downstairs Brain.
- It is crucial that the **natural and logical consequence** matches the original behavior you are wanting to change.
- If you state a consequence will happen if a particular behavior occurs, you must follow through with the consequence when appropriate. Lack of follow-through often results in mistrust, testing limits, and not taking you seriously.
- The given consequence should include a teaching moment for why the behavior was wrong.

If a **natural and logical consequence** involves taking something away (or a loss of a privilege):

- the thing that's taken away should never involve taking away a person or a relationship;*
- the thing that's taken away should be replaced by a person or relational time; and
- the thing that's taken away should not be a physiological or regulation need of the child.

*When a **natural and logical consequence** involves restricting social media or cell phone use, we must recognize and acknowledge that this is the way our tweens and teens connect with most of their friends. And while some of these relationships are superficial, or mere acquaintances, it's likely that a few actually do provide your child with an authentic and emotionally safe space to land. Try to remember this while implementing **structure** and consider ways you might allow for those important relationships to still exist. (Hint: Set aside scheduled, monitored call times each day.)

AN IMPORTANT NOTE ABOUT TIME-INS VS. TIME-OUTS

We regularly practice "time-ins" instead of "time-outs." We never send a child to their room as a consequence or send a child to spend time on their own, as this rarely, if ever, serves as a teaching moment. This is especially important for young children who have not learned to **independently regulate**. Instead, use the opportunity to connect with and co-regulate your child, and once regulated remind them of the expectations.

A young or older teen may be guided to a separate space, to allow everyone an opportunity to calm down (if they are able to independently). However, it is important to reconnect with them as soon as possible to maintain the relationship and process the situation.

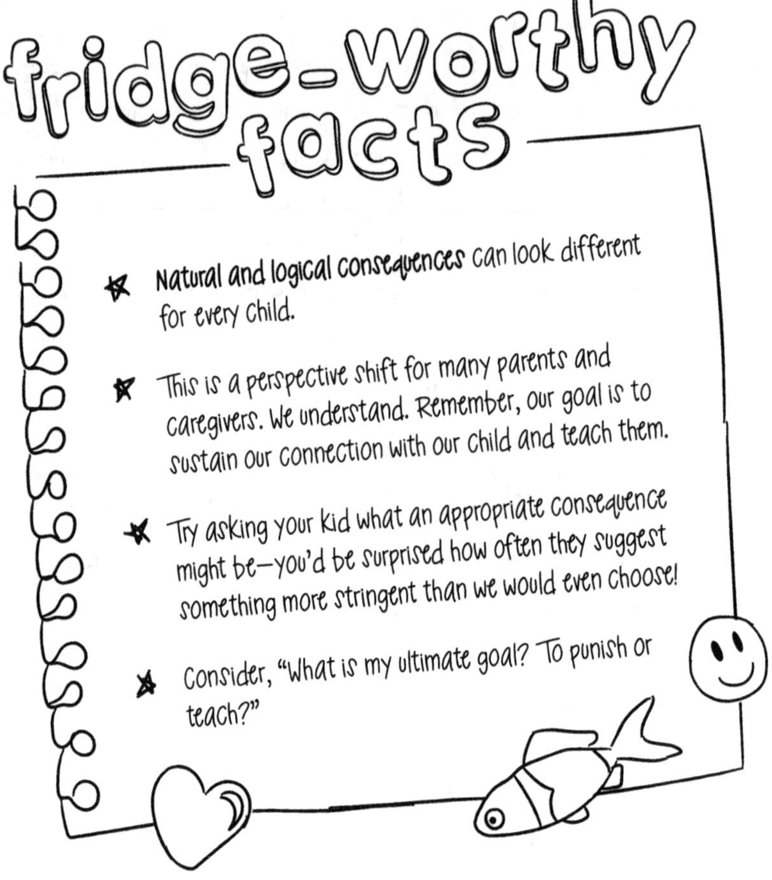

We know that this chapter has introduced a lot of information, some of which might be brand new to you. We don't expect you to know and use it all at once. These are concepts that have taken us years to refine, and we are still always learning, practicing, and getting better at using them. Start small and work your way up, just like we encourage our kids to do.

In conclusion, the DTAP® concepts teach us that presence is more important than perfection. It's about showing up. With curiosity, intention, and patience.

Curious about the need behind the behavior.

Intentional about maintaining a strong relationship.

Patient with the mistakes that both you and your child will continue to make as you navigate not only development but also the digital era.

Part 2

HELPING OUR KIDS BALANCE DISRUPTION AND DEVELOPMENT

IN LIFE'S CO-OP MODE, WE LEVEL UP TOGETHER

If you've been reading up to now, you will hopefully have gathered that it is almost impossible for kids and teens to navigate their development independently. This also applies to their ability to use and manage screens, social media, and technology. This means that we have to take an active role in supporting them, through the good, the bad, and the ugly parts of digital disruption.

Next-Level-Knowledge: If you need a quick refresher about why kids and teens need us to be present with them as they develop and learn, take a look at chapter 1.

We know this is tough. We'd be lying if we said we'd never wished that our kids could just figure things out by themselves. We also understand that it's hard to be vulnerable and take ownership of the part you might play in your child or teen's behavior.

The following important principles are included as a gentle reminder that our own words and actions can determine whether a behavior gets worse or better. We encourage you to keep these in mind as you work to manage challenging behaviors associated with digital disruption:

1. *The caregiver's focus should be on reducing the problematic behavior* and *maintaining a healthy relationship with the child.*

   When our solution is just to punish or stop unwanted behavior, we take away opportunities to teach and build stronger relationships with our children. Remember, you *can, and should,* set limits and consequences (**natural and logical** ones), while being empathic, respectful, and understanding of your child's feelings.

2. *The caregiver must be a part of the solution, and their way of "being" is just as important, if not more so, than their method of "doing."*

   Sometimes it's not about coming up with a solution in the moment; instead, **sitting with our kids in the "yuck"** or the discomfort can be powerful and can help convey empathy and security.

   Remember, just because you are working together it doesn't mean you are letting your child "get away with" unhealthy, or unkind behavior; it is still important to hold them accountable and responsible for their actions—we are just supporting them differently as they learn those skills.

3. *The caregiver should be open to learning from the child or young person.*

   This is especially applicable when supporting adolescents—us older generations may need to accept that we have more to understand before we can collaborate on solutions. It's okay to say, *"Help me to understand"* or *"I'm finding this hard to get, but I understand that it's important to you. Can you tell me more?"*

   Again, this doesn't mean that you have to be okay with your child's behavior choices, but you might have to find a way to compromise as you both figure out the digital world together.

4. *It is always helpful to recognize what part of the brain a child or young person is in when responding to problematic behavior.*

   If we know a child or young person is thinking with their Downstairs Brain (i.e., their survival brain), our response will be much different than if we know they are thinking with their Upstairs Brain. Take a quick look back at chapter 1 if you need a reminder.

   Remember, most often, our children and adolescents are not simply choosing to behave a certain way (we know, sometimes it really seems like they are!); rather, the parts of the brain needed for good choices, logical thinking, and impulse control are either switched off or not fully developed.

5. *It is always helpful to recognize what part of the brain the* caregiver *is in when responding to problematic behavior.*

   The saying "it takes a calm mind to calm a mind" is true for many different and mostly scientific reasons (if you want to know more, learning about mirror neurons is the place to start). Remember, it's more than okay to say, *"I need some time to think about this so that we can come up with a good solution"* and then return to the conversation when everyone is calm.

6. *Strategies may not work every time, all the time. It takes consistency, patience, and reflection to make progress.*

   When we use strategies what we are really doing is helping the brain create new connections through new or different experiences. It takes many repetitions of new or different experiences for a new connection (to the preferred behavior) to develop and become strong, while the old connection (to the unwanted behavior) becomes unused and pruned away.

   Remember when you last learned something new? The chances are you weren't perfect at it right away. So, it's important that we give our kids time and many opportunities to try again, and when you find yourself getting frustrated (which you will!), remind yourself of the other five principles listed above. And don't forget, learning happens faster through **play**, so if you can make learning fun, it's likely that your child will be able to master the changes sooner.

FINALLY

Whenever we are using a strategy or approach with a child, we have found that it is helpful to first ask yourself: *"What is the need behind this behavior?"*

Oftentimes, the answer to this question will guide the way we respond and give us a more accurate understanding of what's really going on.

Most importantly, remember to give yourself grace. Raising children in this digital age is *hard*, and the best thing we can do is *be with* our children as we navigate it together.

HOW TO USE PART 2

First, we recognize that caregiving takes many forms—by parents, grandparents, relatives, foster or adoptive parents, and even teachers. For simplicity, the term "parent" is used throughout to include all who care for children.

We have divided this part of the book into ages and specific tech-related scenarios so that you can easily find support when you need it. However, as with most things that involve development, you will find that there is often overlap and differences between individuals. So, if you can't find the answers that you need in one age group, you'll likely find it in the next one, or the one before.

Throughout the chapters in part 2, you will find words highlighted in bold ; all of these words are concepts that we talked about in part 1, chapter 5.

We can typically manage every scenario in two ways: reactively (in the moment) and proactively (before conflict or dysregulation occurs). We have included explanations for how to do this for every situation.

Finally, we have included our top tech tips for each age group as a go-to resource at the end of each chapter.

And don't forget, these scenarios aren't going to work out perfectly. They can, and will, go wrong, last longer, or sound different. Start small, try one word, phrase, or action and see if anything changes. Take what works for you, and your family. Remember, it's all about relationships, and balance.

Quick Reference Guide

DIGITAL DISRUPTION SCENARIOS

Digital Disruption Scenario #4 Page 181

"My 4-year-old has temper tantrums when we tell him, "No." We began giving him his tablet when this happens and he immediately stops. We know this isn't a good idea but we don't want to keep dealing with his tantrums."

Digital Disruption Scenario #5 Page 187

"My 5-year-old gets to watch his tablet before bedtime while we watch television. Our teenager usually listens to her music in the next room. When we tell him it's bedtime and take him to his bedroom he has a huge screaming and crying meltdown! We don't understand, he just got to do what he wanted for ages."

Digital Disruption Scenario #6 Page 198

"I am so frustrated. I spend so much time fighting with my 6-year-old about when it's time to turn off their tablet. They refuse to give it back to me. I've tried bribing them with all sorts of things. We've set timers and signed contracts. They promise they'll give it up on the time we agree upon, but it always ends up in an hours-long fight. I don't think I can stand to hear 'just one more minute!" another time."

Digital Disruption Scenario #7 Page 204

"Every time my 7-year-old plays a video game, they become destructive and out of control. The longer they play, the worse it gets. I can't get them to listen, they yell and defy me, and they're cursing

or playing really rough with our dog. Taking the video games away just causes a bigger meltdown."

Digital Disruption Scenario #8 — Page 211

"All my 8-year-old wants to do is sit around and play on their tablet all day. I can't remember the last time she went outside or that our family played a game together. Our doctor says she's gained 20 pounds in three months!"

Digital Disruption Scenario #9 — Page 218

"My 9-year-old wants to play video games that I think are too violent for his age and he gets mad when I tell him he can't. Today I found him playing a game he wasn't supposed to be playing. I know I will either end up yelling at him, or just giving in because it's easier than dealing with his anger."

Digital Disruption Scenario #10 — Page 225

"My 10-year-old comes home from school and gets to sit and play video games for two hours while I finish work, before dinner. By dinner time she is so grumpy and we always end up in an argument. I don't know what to do, she loves video games, I thought she would be happy."

Digital Disruption Scenario #11 — Page 236

"My 11-year-old has started to self-harm. I am concerned that some disturbing content he saw online is what triggered his cutting. I know he is hurting and he feels terrible about it."

Digital Disruption Scenario #12 Page 243

"My 12-year-old child spends every weekend glued to social media and then gives me a play-by-play of friends who are at parties or sleepovers to which she was not invited. She used to have fun with us at home, but now she is obsessed with seeing where and with whom she was not included. She spends more time crying on the weekend than not."

Digital Disruption Scenario #13 Page 249

"My 13-year-old daughter complains about being tired and falling asleep in class. I think she is staying up too late, and tonight she said she was going to Facetime her friends to work on homework together. I know if I challenge her it will turn into an argument but she needs her sleep."

Digital Disruption Scenario #14 Page 257

"My daughter hangs out on an online platform and claims she is meeting all kinds of nice gamers from around the country – but we do not know these people, and I read something about extremist groups and predators known to target children for sexual exploitation using this platform to communicate with children."

Digital Disruption Scenario #15 Page 269

"My 15-year-old daughter is constantly on Instagram and TikTok, trying to recreate looks she sees online. Then, she posts photos of herself and starts counting likes and how many followers repost her content. She is never content with the numbers and blames it on her

Chapter 6

STRATEGIES TO MANAGE DIGITAL DISRUPTION IN THE EARLY YEARS (BIRTH–AGE 5)

"Building Balance Before Screen Time Takes Over"

The early years are both amazing and exhausting, for children and adults alike! There is so much development taking place and changes seem to happen in the blink of an eye. Children start crawling one day, but soon enough they're bargaining with you over bedtime treats. It's a beautiful chaos.

While we often wish that time would slow down, sometimes we also wish that our little ones could be just a little more independent so that we could tackle that load of laundry, or even squeeze in a short nap. As parents ourselves, we can truly understand the demands of the early years, and we know that it can be so tempting to just turn on a screen (and it's true, we've all done it) and "zone out." Digital media can easily become a convenient option for caregivers.

The hectic nature of the early years makes it easy to understand why you would use screens to calm a fussy baby, entertain an energetic toddler, or gain fifteen minutes of quiet time to fold laundry. However, it is extremely important for us to remember that infancy and toddlerhood is a time for building strong connections with primary caregivers and family members. Children are learning trust, communication, how to explore their environments, and perhaps most importantly the ability to self-soothe, and digital media use shouldn't crowd out the ability to develop any of these skills. Screen time that is excessive or lacks supervision hinders essential developmental stages such as emotional control and language skills as well as relationship formation. This is why every strategy in this section (and indeed subsequent sections) requires active participation from both the child and their caregiver.

Please note: We do not recommend the use of electronics, screens or digital media during the early years, but we do understand that many parents can feel overwhelmed and helpless during times when their infant is dysregulated, or if they themselves are exhausted, mentally drained and emotionally depleted. We understand that many parents are busy trying to meet the demands of work and family life, and nobody is perfect. What we are looking for is *balance*.

Many of the DTAP® interventions in this chapter will involve connection and building a secure relationship, as well as beginning to establish self-regulation through co-regulation without the use of electronics or screens.

DIGITAL DISRUPTION
SCENARIO #1

"The baby is crying again. I'm exhausted and I really need a break. I could put the TV on; it will calm the baby down and I can sit on the couch and watch."

DTAP® Concepts to Think About		
Downstairs Brain	**Midbrain**	**Upstairs Brain**
Attunement Nurture Window of Tolerance	N/A	N/A

In the moment, here are some things you need to remember, before you respond:

When an infant is crying, they are in their Downstairs Brain, and their distress might cause you to head downstairs too, especially if you are already tired. You need to be aware of your **window of tolerance**, which may be narrowing quickly.

Our goal is to see if we can find ways to maintain connection, **nurture**, and **attunement**, while taking care of our infant's needs *and* our needs.

Here's how we manage this *in the moment*:

- Lay comfortably on the floor next to your baby in the swing and sing gently (be mindful of falling asleep).
- Snuggle on the couch with your baby and recite the alphabet or read a simple book to them.
- Place your infant on an appropriate baby mat with toys, mirrors, and sounds and lie next to them, narrating what you can see.
- Lay comfortably on the floor next to your baby and gently touch their fingers, toes, arms, legs, nose, ears, hair, etc. while describing what you are doing.
- Hold your baby and gently pace back and forth, humming and singing calmly. Take deep breaths. Remember that this will pass. If you need to put your baby in a safe spot, for example their crib, while you take some deeps breaths, this is okay, just stay close by.

Now, here's how we *proactively* prepare for this:

- Create spaces that promote **nurture** and opportunities for connection in your environment, such as playmats, or library books.
- Use **attunement** to stay aware of your infant's needs and try to pay attention to time of day, feeding times, diaper changes, and other environmental stimuli such as noise and light.

DIGITAL DISRUPTION
SCENARIO #2

"My 2-year-old won't sit still unless he has a screen in front of him."

DTAP® Concepts to Think About		
Downstairs Brain	**Midbrain**	**Upstairs Brain**
Attunement Structure and Nurture Window of Tolerance	The Importance of Play	N/A

In the moment, here are some things you need to remember, before you respond:

Infants and children go through some of the biggest physical milestones of their life from ages 0–5, with the locomotion milestone falling somewhere around 1-year of age—so our 2-year-olds *need* to move. Being in the thick of it with my own toddler as I write this, I can truly tell you that I understand needing a break or wanting to get a chore done but "they just won't stop moving!" It's hard and it takes **attunement** to pay attention to their movement needs.

Our toddlers are learning through their environments and the people in them, and we have to remember the **importance of play**: Luckily, with a little **structure** any activity can be turned into something fun and playful, especially if we can also incorporate **nurture**.

Finally, we can use our knowledge about the **window of tolerance** to help our children (and their bodies) learn how to have high-energy, busy time and low-energy, quiet time. However, they will only be able to manage each type of activity for short periods of time before they again reach the edge of their **window of tolerance** (check back to chapter 5 if you need a reminder of this).

Here's what to say to manage this *in the moment*:

Parent: Wow your body is so wiggly! We are going to play a game that will help you get some of those wiggles out!

(You are using **attunement**.)

"Let's pretend to be animals and have races up and down the hall! First, I am going to be a fast cheetah. Can you go as fast as me?!"

(You are setting **structure** and remembering the **importance of play**.)

"High five, that was awesome!"

(You are providing **nurture**.)

"Okay, now let's go really slowly but loudly and stomp like an elephant, stomp, stomp, stomp."

"High five, you did great!"

Now in a quieter voice: *"Okay, now let's go slow and quiet like a snail, shhh."*

"High five!" (Do this quietly.)

Now in a whisper: *"Okay, now let's go as quietly as possible and slow like a snake."*

Whisper again: *"Wow that was so good."*

You can repeat this activity two or three times, or as many times as your child will play for!

Then we will encourage our child to settle into a quieter activity:

"Phew, my body is feeling tired now. Let's get a drink of water and then we are going to look at some books."

(You are using **attunement**.)

"I'm going to fold some laundry now while you look at some books, or you can play with your toys, or you can help me."

Remember, toddlers love to help! If there is a way you can engage them in the task you have to get done, this would be a great option, perhaps they can put all the socks into a basket for you.

Or if your toddler is playing, be sure to engage them in a conversation or narrate what they are doing while you do your chore(s); this will help keep them interested in what they are doing.

Now, here's how we *proactively* reinforce this:

Parent: Good morning, my little wiggle monster! I am so glad to see you! Wiggle on over here so I can give you the biggest squeeze!

(You are using **attunement** and **nurture** to provide deep pressure, which can aid in focus and regulation.)

"Oh, that was such a good squeeze! Before we do anything else, we've got to get our morning wiggles out. I bet they can help us decide what's for breakfast today! Hop if you want eggs and stomp if you want pancakes!"

(You are setting **structure** through **play**.)

"Look at those hops! Eggs sound great! Now, kick like a ninja if you want blueberries or spin like a ballerina if you want grapes!"

"Those are the best ninja kicks I've ever seen! Blueberries it is. Okay, one more, (say this quietly) *walk to me like a quiet mouse if you want milk or slither like a silent snake to me if you want water."*

(Keep your voice calm and quiet.)

"Water sounds perfect! I think I need some too. High five. Come on over here and look at these cups. Would you like the blue cup or the green cup?"

(You are providing **nurture** and setting **structure** through choices.)

"Great choice. Would you like to play with blocks, or help make breakfast?"

When you use activities like the ones in the example on a daily basis you are helping your toddler get their movement needs met proactively, which can make things a little easier when they have to sit still for a while.

We won't lie though, this can be an exhausting age and our little ones definitely keep us on our toes. However, the more proactive you can be in prompting movement (when you have the energy!), the more likely they are to be still when we need them to be.

And finally, there's nothing wrong with using a video or two to help them move. Just try not to make this the norm, and definitely don't leave them to watch by themselves. (This is when "zone-out mode" is activated!)

DIGITAL DISRUPTION
SCENARIO #3

"My 3-year-old struggles when we have to go on a car ride. She whines and cries. When she gets to look at a tablet she is fine, I don't know what else to do; it's just easier."

DTAP® Concepts to Think About		
Downstairs Brain	**Midbrain**	**Upstairs Brain**
Attunement Structure and Nurture Window of Tolerance	The Importance of Play Sitting in the Yuck	N/A

In the moment, here are some things you need to remember, before you respond:

Much like our last example, movement and regulation play a big part in this scenario. If you have a toddler you will know that they don't really enjoy being strapped in, or stuck in one position for any period of time (and to be honest, who does?). This means that car rides, high chairs, and shopping carts are all fair game for a meltdown. When we use a screen to avoid this, we are losing the opportunity to teach our child helpful ways to remain in their **window of tolerance**, as well as the chance for connection and learning.

Thankfully, this situation is one with lots of options. We can show empathy through our use of **nurture**, or by **sitting in the yuck**, and we can maintain expectations through our use of **structure**. We can also use **play** to engage our child in fun activities while also building our relationship with them. Our ultimate goal is co-regulation.

We are going to give you an example of how to manage in the moment, but we want you to know that it's not always going to be perfect, and occasionally the only thing that's going to work is Elmo's World. It's okay. Let's just try to make that the last resort.

You've got this!

(P.S. If you can sit next to your child, this will help make co-regulation easier.)

Here's what to say to manage this *in the moment*:

Parent: Ugh, I know you hate car rides, I understand, sometimes I do too. I'm right here.

(Proximity = **nurture**.)

Child whines and cries: I want my tablet!

"I know you do! I also know your body really wants to be moving. Let's move our different body parts by singing 'Heads, Shoulders, Knees, and Toes.'"

(Here, you are using **attunement**, **play**, and **structure**.)

"No!"

"I'm sorry you feel mad right now. It's hard being stuck in your seat, I get it. Phew, I need to take a deep breath!"

(This is **sitting in the yuck**.)

"Okay, I'm ready to sing! Let's start with one of your favorites!"

(Engage your child in some age-appropriate songs. This could be a good time to put some music on. This is a good use of **play** and **attunement**.)

"Okay, that was fun! Let's see what we can see outside our window."

(Engage your child in looking out of the window.)

"Wow! So much to see outside. I wonder what things we can see inside the car?"

Here are some other ways to engage your child in a vehicle or situation that requires them to be strapped in. Remember, your child will be more engaged for longer if you *interact* with them while they are using items listed below:

- Nursery rhymes or songs
- Books
- Drawing/Coloring
- Magnets
- Toys
- Snacks
- Drinks
- Building blocks

- Stickers
- Noticing things outside

Now, here's how we *proactively* reinforce this:

Parent: Okay, we are going to be taking a car ride this afternoon. I know it's sometimes hard for us when we do that and you don't enjoy it. Let's choose two toys to take for you to play with.

Child: Okay.

"Also, we can play your favorite songs while we ride."

"Can we listen to 'Baby Shark'?"

"Sure!"

"We will have snacks too. All of these things will help us feel better while we are driving."

Being cooped up is hard for everyone, road trips especially, as unexpected things often happen—traffic jams, toilet accidents, or even freak weather incidents (yep, living in the Midwest doing a road trip during tornado season = not fun). But the more prepared you can be, the better for you and your kids, and the less likely you are to have to use electronics for the whole ride.

DIGITAL DISRUPTION SCENARIO #4

"My 4-year-old has temper tantrums when we tell him no. We began giving him his tablet when this happens and he immediately stops. We know this isn't a good idea, but we don't want to keep dealing with his tantrums."

DTAP® Concepts to Think About		
Downstairs Brain	**Midbrain**	**Upstairs Brain**
Attunement Structure and Nurture Window of Tolerance	The Importance of Play Sitting in the Yuck	Natural and Logical Consequences

In the moment, here are some things you need to remember, before you respond:

When your 4-year-old has a temper tantrum they are in their Downstairs Brain, and it is easy for you to join them down there, especially if this is tantrum number 5 and it's only 9:00 a.m. You need to be aware of your **window of tolerance**, and your child's, by using **attunement**.

A 4-year-old is learning about cause and effect: If I do A, I get B. As a result, they will adjust their behavior to get the response they want (and have come to expect). In this scenario, our

4-year-old may have already come to expect the tablet when they are upset.

Additionally, by giving our child a screen whenever they are upset after being told no, we are teaching them that tantrums = tablets and that's the only way to calm down. It's important to teach our kids to tolerate big feelings and realistic ways to regulate so that they can cope when technology isn't available (at school for example).

To make it through this tech tantrum we will need to use **structure** and **nurture** to manage challenging behavior and be prepared to **sit in the yuck** with our child as they learn to regulate without the use of a screen. We might even have to follow through with **natural and logical consequences**, depending on our child's behaviors during this time.

Here's what to say to manage this *in the moment*:

Parent: You can't go outside right now.

Child (starting to get angry): I want to go outside!

(Child goes to grab the tablet.)

Parent: We aren't going to use the tablet right now.

(You are setting **structure**.)

Child (screaming): Give me the tablet!

"Wow, you are really mad right now."

(Get down on their level if it's safe to do so. Increasing proximity is a way of providing **nurture**.)

"I hate you! I want the tablet!"

"That's a big feeling. I'm going to take some deep breaths."

(Here you are paying attention to your **window of tolerance** and **sitting in the yuck**.)

(Take some deep breaths in front of your child.)

Child continues to scream and starts to run around throwing toys.

"I need you to be safe. You can throw the cushion."

(**Note:** This behavior will be addressed, but not in the moment when the child is in their Downstairs Brain.)

Parent remains calm and takes deep breaths.

"I'm here when you're ready."

(Child's movement begins to slow down.)

"I want to go outside."

"I know, you love playing outside. Sit with me and take some deep breaths."

"Okay."

(Provide **nurture** to your child and continue to model taking deep breaths.)

When you are sure your child is out of their Downstairs Brain, you can begin to address what just happened.

Quick Tip(s): To check if your child is out of their Downstairs Brain you can ask them a simple question such as "Hey, do you remember what the weather was like yesterday?" If they can give a clear, correct answer, they are out of their Downstairs Brain.

"Phew, those were some big feelings we just had."

"Yeah."

"I know you really wanted to go outside, I'm sorry that the weather is so bad right now and we can't go outside."

(Child is quiet.)

"It's okay to feel mad when things don't go your way, but it's never okay to throw things. Do you know why?"

"Because they might break?"

"Yes, or you might hurt yourself or someone else."

(Here, you are introducing **natural and logical consequences**.)

"You said I could throw a cushion."

"You're right, I did. We need to figure out another way to get our big feelings out that doesn't involve throwing."

"I could use the tablet."

"I like that you are thinking about it. But you won't always be able to use your tablet. What happens at school when you have big feelings?"

"We smell the flowers and blow out the candles."

"Oh wow, show me!"

(Child mimics taking deep breaths.)

"That's a great idea! Because you don't have a tablet at school, right?"

"No."

"How about we try that at home next time we have big feelings?"

"Okay."

Now, here's how we *proactively* reinforce this:

Parent: Hey, kiddo, what do we usually do before we go to bed?

(The use of a simple question is a great way to check that your child is calm and not in their Downstairs Brain).

Child: We take a bath and brush our teeth.

"That's right! Hey, do you remember yesterday when you wanted to go outside and we couldn't because it was raining?"

"Yeah."

"Yeah, I know, you were sad. So sad that you wanted to throw your toys. Do you remember why we can't throw things?"

"Because we can break our toys."

"That's right and we can hurt people. Mommy used to give you your tablet when you had big feelings but we have to smell the flowers and blow out the candles instead now."

"Can you show me again how to do that trick?"

Child demonstrates.

"Oh wow! That's so neat! Am I doing it right?"

(Remember the **importance of play**, try doing it wrong so your child has to correct you.)

Child laughs. "No, Mama, you're doing it wrong, watch me!"

"Oh, silly me! Let me try again."

Practice more deep breaths with your child.

"Okay, so next time I notice you having big feelings, I will say, we need to smell the flowers and blow out the candles! Okay?"

"Okay."

We get that this can all sound a bit "too good to be true." And you're right, it won't always work out like this: The tantrum will last longer, tears will be spilled (maybe even yours), and we might say things we don't mean in the moment. If you're not ready to stop using the tablet right away, try reducing the amount of time your child gets to use it a little bit at a time, while also teaching them a new way to handle the big NO.

DIGITAL DISRUPTION SCENARIO #5

My 5-year-old gets to watch his tablet before bedtime while we watch television. Our teenager usually listens to her music in the next room. When we tell him it's time for bed and take him to his room he has a huge screaming and crying meltdown! We don't understand; he just got to do what he wanted for ages."

DTAP® Concepts to Think About		
Downstairs Brain	**Midbrain**	**Upstairs Brain**
Attunement Structure and Nurture Window of Tolerance	Verbal Responses Sitting in the Yuck	N/A

In the moment, here are some things you need to remember, before you respond:

Your child is in their Downstairs Brain (and it's likely you might be too) because it's the end of the day and everyone is tired. Not only that but your child is also at the edge of their **window of tolerance**. Remember, young children find it difficult to recognize and regulate sensory input (such as sound, music, and lights) so that they don't become overloaded, or understimulated. When this happens, they can move out of their **window of tolerance** resulting in the meltdown described above. In this specific situation, the child's sensory system cannot manage the drastic shift from bright, noisy, and fast, to quiet and dark. It will take **attunement** to help them learn these skills over time.

To avoid this type of sensory meltdown in the future you will need to slowly adjust your child's bedtime routine to something quieter, with less stimulation.

If this has become a typical evening routine, it's going to be hard for your child to adjust right away. Instead, you will have to use a balance of **structure** and **nurture** and be prepared to **sit in the yuck** and use **verbal responses** to help your little one cope as you gradually introduce them to the new expectations.

Here's what to say to manage this *in the moment*:

First, take a deep breath. Nothing good will come from everyone being in their Downstairs Brain. Trust us, we've been there.

Before using any words, we need to co-regulate, do this by:

- taking lots of deep breaths,
- picking up and soothing your child (if they are being physically safe), or
- being with them until they have calmed down.

Once their breathing has become more regular and they are moving less, then you can talk. Keep the words you use short and simple. Remember, all we are doing right now is reconnecting with our child and helping them go to bed calmly. Tomorrow we will address the problem.

Parent: I know this is hard. It's bedtime now.

Child cries softly.

"I'm sorry this has been so hard for you."

(Here you are **sitting in the yuck** with your child.)

"We need to do some things differently from now on to help us when it's bedtime."

(You are using **attunement** and verbalizing that he is outside of his **window of tolerance**.)

"Like what?"

"We will talk tomorrow when we aren't tired, okay?"

"Okay."

Now, here's how we *proactively* reinforce and prepare for this:

At breakfast time (it's important that you give your child enough time to prepare for the changes ahead):

Parent: Hey, kiddo, bedtimes have been kinda hard, haven't they?

Child: I hate going to bed.

"Oh wow, thanks for telling me that. I'm sorry you don't like it. I've noticed that you have to go into a dark, quiet room after being in the living room where it's loud and bright. That makes it hard for our brains and bodies to relax. I wonder if that might be why you hate going to bed so much."

(Your **verbal responses** are helping your child connect what's going on outside his body with what's happening inside his body.)

"No, I want to play on my tablet; I don't want to go to bed."

"Hmm. Okay, thanks for letting me know that. Today we are going to try something different. After dinner you can have your tablet like you usually do, but at seven thirty we are all going to stop with our electronics and have some quiet time."

(You are setting **structure**.)

"You and Dad too?"

"Yes, me and Dad too. You can choose what we do for thirty minutes before bed. We can read books or play with our toys. We will also take a bath and put pjs on."

"But I want to play on my tablet."

"I know, it will be hard to begin with, but we will be there to help you. We could even get a few new books at the library or store if you want."

(You are providing **nurture** and **sitting in the yuck** by being with your child through this.)

"Can we get the books with stickers in them?"

"Sure."

Is your child still going to struggle when it comes to quiet time? Absolutely. We're not here to sugar-coat anything. If this happens, go back to managing in the moment and revisit again, as many times as needed. But with consistent and predictable **structure** and **nurture**, it *will* become easier.

TOP TECH TIPS FOR THE EARLY YEARS

1. Keep screens and tech use to an absolute minimum, none if you can manage it.
2. If you do incorporate screen time, make sure you are engaging with your child at the same time—ask them questions, dance, sing a song with them, narrate what you see on the screen.
3. Keep tech time away from bedtime—sensory overload will make it much harder for your little one to fall asleep and stay asleep. This means screens, loud televisions, loud music, and lights need to be turned down or reduced.
4. Put your own screen down too. Little kids need their caregivers for almost every physical need, and emotional ones too, and they need to be seen and heard; otherwise, they will start behaving in ways that *make* you see and hear them.
5. Balance screen time with outside time, physical activity, toys, games, books, and learning opportunities.

QUICK SWAPS TO TRY

Instead of this...	Try this...
Background TV all day	Gentle music, children's nursery rhymes, or silence with frequent playful interaction.
Screen to soothe a tantrum	Rocking, singing, a change of scenery, or just sitting near your child while they feel big feelings.
Videos during meals	Talking about what's on your plates, asking questions, or silly storytelling.
Tablet time in the car	Playing I Spy, singing together, listening to stories, or spotting things outside.

These small strategies over time can have big impacts.

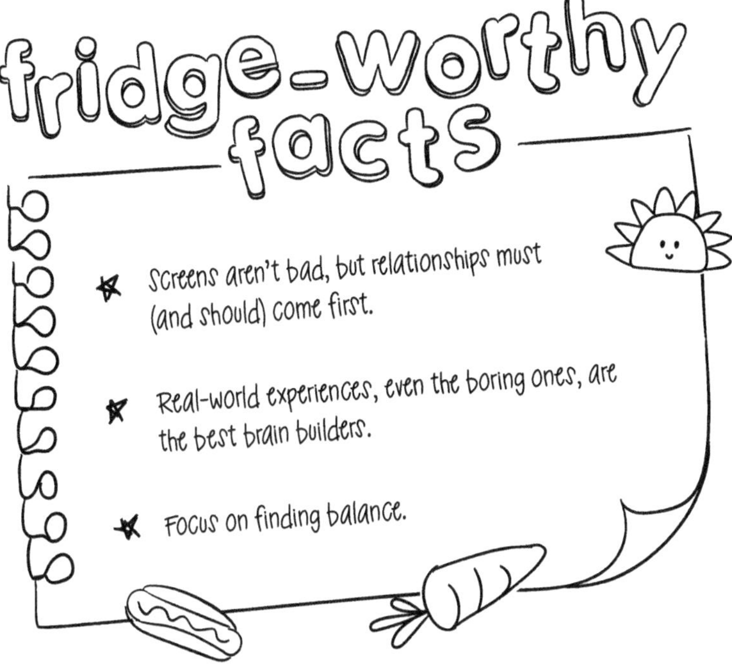

Be prepared for tantrums. Remember your own **window of tolerance** and what helps you get regulated. The more we can be with our child during challenges, the less frequent they will become. We've been there. You've got this.

Chapter 7

STRATEGIES TO MANAGE DIGITAL DISRUPTION IN THE CHILDHOOD YEARS (AGE 5–AGE 10)

"Helping Children Stay Grounded as the Digital World Expands"

The school age of development can be such an exciting time: Kids are typically starting school, beginning to navigate friendships, asking you for help with homework (maybe not quite so exciting!), playing on sports teams, or participating in extracurriculars. They're learning big things, like how to control impulses, express their feelings, and get along with peers. That little kid you once knew might seem to have become a "big kid" overnight!

Challenging behaviors relating to digital disruption will occur and when they do it's important that we keep relationships and connection in mind. And just to remind you, that doesn't mean getting rid of **natural and logical consequences**.

You'll notice that we do not accomplish this goal by banning the use of electronics completely; instead, we'll give you suggestions on how to set limits and **structure** surrounding technology usage in the home in a relationship-focused manner, using **nurture**. While it is an option, giving your child no access to electronics is often not realistic: Tablets, phones, handheld gaming systems, television, and video games are commonplace in schools and homes. Our bigger goal is to teach you how to include electronics without them becoming the only focus of time at home or an interference to important tasks and activities.

Many of the DTAP® interventions you're going to read about will involve **attunement** and widening a child's **window of tolerance** and ability to remain regulated for longer periods of time without electronics. We will also explore ways to increase a child's tolerance for managing overwhelming emotions when it's time to stop using an electronic device, with a balance of **structure** and **nurture,** and **verbal responses** to wonder about, validate, and honor their experiences. We often find if we remove electronics "cold turkey," this causes more behavior challenges and conflict between caregivers and children than if we slowly taper time on screens over time. We know it can be daunting to set limits when it comes to electronics—we've experienced our fair share of tech tantrums and trust us, nobody wants to deal with the meltdowns—but several of our interventions include choices and compromise, which can help you feel confident and give your child an appropriate dose of power, making everyone's lives a little less stressful.

Much like our previous chapter, you'll see that when we are asking a child to participate in an activity other than screen time, we're often doing an activity with them instead of asking them to do something else independently. While **independent regulation**—"Hey, time to get off your tablet. Go do something else for a while"—is the goal, children (and even teens) can't always do this until an adult teaches them. This is done through repeated relational experiences, such as: "Hey, ten more minutes on your tablet and then you and I have some basketball to play!"

DIGITAL DISRUPTION
SCENARIO #6

"I am so frustrated. I spend so much time fighting with my 6-year-old about when it's time to turn off their tablet. They refuse to give it back to me. I've tried bribing them with all sorts of things. We've set timers and signed contracts. They promise they'll give it up at the time we agreed upon, but it always ends up in an hour's-long fight. I don't think I can stand to hear, 'Just one more minute!' another time."

DTAP® Concepts to Think About		
Downstairs Brain	**Midbrain**	**Upstairs Brain**
Attunement Structure and Nurture Window of Tolerance	Verbal Responses Sitting in the Yuck	Problem-Solving and Abstract Thinking

In the moment, here are some things you need to remember, before you respond:

Trying to reason with an emotional 6-year-old is near on impossible; they are most definitely in their Downstairs Brain. Instead, focus on your own **window of tolerance** and decide when the best time to address this would be for you, and for your child.

Our **verbal responses** are going to be important as we help our child accept the changes that need to happen to avoid meltdowns like this in the future. We might have to **sit in the yuck** with them as they accept that they can no longer have their tablet all the time.

6-year-olds are only capable of simple **problem-solving** and **abstract thinking**. Therefore, whatever **structure** we put in place has to be simple, and balanced with **nurture**.

Instead of attempting to address this already stressful situation in the moment, we are going to give you an example of how to proactively implement gradual changes to help make the transition easier for everyone.

However, just in case you've found yourself in the middle of a power struggle with a miniature version of yourself . . .

Here's what to say to manage this *in the moment*:

Child (yelling): I'm not giving it to you! It's my tablet!

Parent: Wow, those are big feelings! I have some of those too right now.

(This is a great example of **verbal responses**.)

"I'm playing my game, and you can't stop me."

"Thanks for letting me know. I need a break and a drink of water."

(It's perfectly okay to tap out.)

Note: We are not going to address the issue yet. Everyone (including us) needs time to calm down before we can attempt

problem-solving and abstract thinking. Remember, this doesn't mean they are "getting away with" this behavior; it means we are choosing to handle it in a smarter, more effective way.

Now, here's how we *proactively* prepare for, and reinforce this:

In the evening, at bedtime, preferably when you are providing **nurture** to your child . . .

Parent: Hey, buddy, we are having a really hard time figuring out the tablet, aren't we? We seem to get mad about it a lot.

Child: Yeah.

"Well, after school tomorrow we are going to talk about what we need to do together to make sure we don't get so mad, is that okay?"

"Okay."

Provide a reminder before school . . .

"Hey, kiddo, don't forget that after school we are going to have a chat about those big feelings we both have about the tablet."

After school . . .

"Hey, honey, now that you've had your snack we need to talk about our big feelings we both have about the tablet."

(Talking after a snack is a good demonstration of **attunement** to your child's needs.)

"I like my tablet."

"Yes, I've noticed that! I've also noticed that you and I get mad with each other sometimes when you are playing with it. Have you noticed that?"

"Yeah sometimes."

"I wonder if you feel mad when I ask to you turn it off?"

(Great use of **verbal responses** to help your child recognize their feelings.)

"Uh-huh."

"Do you know why you feel mad?"

"Because I like playing my game."

"Yes, it looks like a lot of fun! Sometimes you get so into it I wonder if you can hear me?"

"Sometimes I don't."

"Hmm, that's good to know!"

"It seems like it's really hard for you when I tell you it's time to stop playing, and we've tried some other ways to help us not feel mad but they don't seem to help. Today we are going to try something new. We are going to practice stopping activities at the right time. We're going to start practicing using some games that we know!"

(You will need to be enthusiastic about this!)

"C'mon, we are going to play musical statues!"

(Playing games such as this with your child is not only fun, but it teaches them how to follow rules. Continue to engage your child in activities that help them practice stopping a fun activity when they are told to.)

"Now, today we are going to do exactly what we do when we play musical statues, but we are going to practice it with our tablet. You get to play on the tablet, but when the music stops, you have to pass it to me! Let's try it!"

"Okay."

Over time, increase the length of time your child has the tablet for. If they can demonstrate success, it might be time to introduce a timer again.

Of course your child is going to find this easier to do while you are playing with them (we've been around 6-year-olds a time or two), but this practice will not only start creating new brain connections (ones that associate giving the tablet up with fun) but will provide you with a way to scaffold your conversation next time your child gets to play on their tablet.

It will sound something like this:

"Hey kiddo, I want you to be able to have tablet time today. We have been practicing giving it up at the right time and you seem to be doing good at it. How about we give it a real try today? But we will keep it short. Would you like 10 minutes or 15 minutes of tablet time?"

"15 minutes!"

"Okay, great choice. When you have 5 minutes left I'm going to come and remind you."

"Okay."

This is a scenario that is going to take time, patience, and follow-through to fix. Especially follow-through. You will want to give in and give your child the tablet because it's easier in the moment. But remember, it's harder in the long run.

We know you can do this!

DIGITAL DISRUPTION
SCENARIO #7

"Every time my 7-year-old plays a video game, they become destructive and out of control. The longer they play, the worse it gets. I can't get them to listen, they yell and defy me, and they're cursing or playing really rough with our dog. Taking the video games away just causes a bigger meltdown."

DTAP® Concepts to Think About		
Downstairs Brain	**Midbrain**	**Upstairs Brain**
Attunement Structure and Nurture Window of Tolerance	Verbal Responses Sitting in the Yuck	Problem-Solving and Abstract Thinking Natural and Logical Consequences

In the moment, here are some things you need to remember, before you respond:

Your child's brain is currently overstimulated, they are outside of their **window of tolerance**, and they may even be in their Downstairs Brain. You need to use **attunement** to determine this.

Your child will not be able to take in a lot of information at this moment, so you need to keep the **structure** you're about to set simple.

You already know that taking away the video game is going to be hard for your child, so you will need to **sit in the yuck** with them. If their behavior is hurtful, you may need to enforce **natural and logical consequences**.

Once you and your child have calmed down, you can use **verbal responses** to help them feel better and begin to use **problem-solving and abstract thinking** to avoid the same thing happening again.

Here's what to say to manage this *in the moment*:

Parent: It looks like this game is giving you some big feelings. In three minutes, we are going to pause it and take a break. I'm setting a timer now.

(You are **attuned** and you are setting **structure**.)

Child: I'm not stopping!

"I understand, our timer is set."

"Our timer is going off, so you can choose to pause it or you can choose to lose it for the day. Which do you choose?"

"I'm not pausing it!"

"You can choose or I can choose for you. Which would you like?"

(Here you are setting **structure**. Ideally, we would like our child to pause the game, but if they refuse, you need to stay calm and follow through.)

"I see you are finding it hard to make a choice, so you won't be able to play anymore today."

Child throws the console, it breaks, and they start screaming.

"I know you're feeling really mad with me right now, and that's okay. We will take a break together."

(Remain calm and if safe to do so try to sit down near your child (provide **nurture**) and model taking deep breaths. Allow the child to pace and yell, as long as they are not hurting themselves, others, or property. This is **sitting in the yuck**.)

"I'm here when you're ready to take a deep breath."

(Continue to allow the child to move and make noise if they need to. This is their attempt to self-regulate and get back within their **window of tolerance**. When their movements subside and their breathing is settled, you can begin to address what just happened.)

"Phew, those were some big feelings you just had."

"It was all your fault."

"I'm sorry you feel that way. It doesn't feel good when things don't go our way."

"I was fine until you made me stop."

"Can you tell me what 'fine' looks and sounds like for you?"

"I was playing my game."

"I wonder what your body and voice were doing while you were playing your game?"

(This is a great way of using **verbal responses** to gently explore your child's behaviors.)

"Well, I was just yelling at the game because it was making me lose."

"Oh yes, I remember, you were yelling; it was quite loud. You don't normally yell that loud, do you?"

"No."

"That's right, I don't ever have to tell you to stop yelling, unless you are playing your video game. I wonder what that's about?"

"I just get so mad at it because it makes me lose."

"So, when you start losing the game, you start getting mad. Do I have that correct?"

(Your **verbal responses** are helping to identify and validate their feelings for them.)

"Yes."

"Hmm okay, and when you get mad, you start yelling loudly and I noticed you throw things and sometimes you even play rough with the dog. What do you think about that?"

(This is expressing that you see and hear what is happening, you are conveying **attunement**.)

"Yes, that happens."

"I wonder if it would be helpful to take a break before that happens?"

(You can now introduce **problem-solving and abstract thinking** to your child to help them find solutions.)

"I think so."

"Okay, I can help you figure that out."

Now you can work collaboratively with your child to identify ways to help them know it's time to take a break. Talk about their voice, body language, and words they are using. Then identify a method for letting them know, for example, a timer, a secret verbal or physical signal, a gentle touch, etc. The most important thing is that you find a method that you and your child can agree on (but remember, it can always be changed if it doesn't work).

You will also need to address the issue of your child throwing the console:

"It looks like we might not be able to play on that game for a while until the console is fixed. This will give us some time to practice taking breaks."

"But I want to play my game!"

(It can be tempting to give in at this point because your child is sad, but it's important that you follow through with the **natural and logical consequence** that has occurred to reinforce the learning that is happening.)

"I understand that your game is important to you, but right now it is broken and we will have to get it fixed."

"It's not fair!"

"I know you feel sad about that, but we can find some other activities to do together until it is fixed again."

"I don't want to do something else!"

"It sounds like you aren't ready yet. That's okay. I'll be here when you want to do a different activity with me."

(Give your child some time to calm down but stay close by. You could even begin doing a fun activity and see if they join in.)

Now, here's how we *proactively* reinforce this . . .

Child: Can I play my video game?

Parent: Yes, but do you remember what we need to do if you start to feel mad or if I notice you starting to look mad?

(This question reminds your child of how to use **problem-solving** skills in times when they feel dysregulated.)

"You're going to say, 'Calm in three minutes' and start a timer."

"That's right, and remember, it's just a break. You can start playing again if and when you are feeling calm, sound good?"

(By providing this explanation you are helping your child develop their **abstract thinking** skills.)

"Yes."

"And what happens if we can't calm down?"

"We have to find something else to do."

(Now your child is demonstrating their understanding of **natural and logical consequences.**)

"Okay, great, you've got it!"

It's important to remember that just because you allow your child to do another activity that they enjoy, you aren't letting them "get away with" the unwanted behavior (throwing the console). The **natural and logical consequence** is that they can't play with it anymore, and it's not necessary for us to enforce additional punishments on top of that. If you need a reminder about **natural and logical consequences**, turn to Chapter 5, page 144.

DIGITAL DISRUPTION
SCENARIO #8

"All my 8-year-old wants to do is sit around and play on their tablet all day. I can't remember the last time she went outside or that our family played a game together. Our doctor says she's gained twenty pounds in three months! When we try to encourage her to play outside with us she yells and gets angry."

DTAP® Concepts to Think About		
Downstairs Brain	**Midbrain**	**Upstairs Brain**
Structure and Nurture Window of Tolerance	The Importance of Play Verbal Responses Sitting in the Yuck	Problem-Solving and Abstract Thinking

In the moment, here are some things you need to remember, before you respond:

While your child appears calm on the outside, the likelihood is that she is *too calm* and outside of her **window of tolerance**, in a hypo-aroused state—tired, grumpy, low-key—which is not helpful when we want her to be active. It's also likely that when we disturb her from that state, she will jump into her Downstairs Brain because we are asking her brain and body to make a BIG shift.

Given that we know this, we don't recommend trying to cajole your child into physical activity without any prior warning. If you do try, sometimes using **play** can help avoid a meltdown but it's likely you will need to use **structure** and **nurture** to help her regulate, and provide **verbal responses** that validate her experiences—we may even have to **sit in the yuck** with her for a while. We will explore that scenario below.

A better idea would be to set aside a time when you know your child is regulated and in their Upstairs Brain. This is a sensitive topic: Your 8-year-old is starting to learn more about her identity and her body image, and may even be influenced by images she sees online or on her video games.

Here's what to say to manage this *in the moment*:

"Hey, honey, would you like to come and play outside with us?"

"No! I'm playing my game!"

"Okay, but it's a nice day outside."

"I told you no! I don't want to!"

"Okay, thanks for letting me know. I need your attention for a minute though."

"Ugh."

(Wait until the tablet has been paused or put down.)

"After dinner tonight we are going to have a chat about how much time we spend on our electronics and how much time we spend doing other things."

"Fine."

Remember, this is not the time to get "into it" with your child. We need to set **structure** and talk about expectations when everyone is calm and has access to their Upstairs Brain. Choose your battles wisely.

Some things to think about before we talk with our child:

1. What Does Your Child Like To Do?

Before electronics took over, what did your child like to do to move their bodies? Jump, run, swim, ride their bike, dance, swing, go for a walk, or ride their scooter? These would be things we would lean into participating in *with* the child when asking them to move their bodies.

2. Start Small

We must recognize that asking a child to spend more time moving and less time playing on an electronic device is a big deal for an 8-year-old, so we need to introduce this in small doses throughout the day. This could look like starting with a ten-minute walk in the morning before any screen time or ten minutes of physical activity in the living room (maybe a dance party with her favorite music). In a week or two, movement time could move up to two, or even three, times per day.

3. "First This, Then That"

To set everybody up for success, plan to move your bodies *before* the child has time on their electronic device. This way, you're not already having to fight with your child to get them away from the screen. "First this, then that" statements can help let your child know what's coming and set **structure** around physical activity and technology: *"I know you want to play your video game. First, we're going to jump on the trampoline for ten minutes, then you can play."*

4. Lean Into Resistance

It's almost unavoidable to encounter some groans, "no's," or even, "you're out of your mind" from kids surrounding this particular issue. Lean into it! When adults are asked to exercise, diet, or add in an inconvenient routine for our health, we often give some flack too! **Verbal responses**, **play**, **structure** and **nurture**, and **sitting in the yuck** are our go-to suggestions for managing in the moment:

a. "But I don't want to go outside! This is so stupid!"
   "Honey, I hear you. I know this isn't something you want to do. Our bodies need movement to help us grow and be strong. C'mon, we'll make it fun."

b. "Dad, I don't get it. Why are you making me do this now?" *"I understand why you're confused. And maybe a little frustrated. Our bodies need movement to grow and be strong. We need to make some changes to be healthy. Let's do it together."*

c. "Uh, I'm not doing that. No way." *"Thanks for letting me know you aren't ready yet! Let me know when you are."* "Okay. Where's my tablet?" *"Whoops. Maybe I wasn't clear. You and I are going to do some physical activity first for ten minutes before tablet time can take place. You can choose! Would you like to jump on the trampoline or take a walk?"* "What?! That is so stupid. Forget it." *"Okay! I'll be in the living room when you are ready to choose!"*

Notice when resistance takes place, the adult stays neutral while also not shying away from the resistance the child is giving. We are okay with (and even want to encourage) our child having a choice but we will maintain our **structure** about what needs to happen first before time on electronics begins. We're not feeding into it, but we're also not avoiding it.

Okay, let's see how this all plays out in a real-time conversation when we *proactively* reinforce this:

Parent: Okay, kiddo, let's chat a little bit about our electronics. What do you love doing on your tablet?

(We are connecting with our child first, instead of diving right into the hard stuff. You get bonus points if you can sit close to your child and provide **nurture**.)

Child: I've been playing this matching game and I am at a super high level; it's so fun!

"Oh yeah? It feels good when we get good at something, right?"

"Yeah."

"Do you remember when you learned how to ride your bike and you got so good at that?"

"Yeah, it didn't take me long."

"You're right! I loved doing bike rides with you."

"Yeah, me too."

"And we used to have dance parties in the living room, do you remember?"

"Yeah, except you can't dance very well."

"Haha, yes you're right. It seems like we don't do any of those things anymore. All of us spend a lot of time looking at our screens; I miss spending that time with you."

(Using humor brings **playfulness** to the interaction and makes the difficult conversation easier to have.)

"Yeah."

"I have an idea. Tell me what you think about it. How about we try to do two bike rides each week, before we play on our electronics? We can pick nice days, not when it's raining."

(This is a good example of setting **structure** and beginning to **problem-solve.**)

"I guess."

"Yeah I know it will be a bit different, but I know I need to get my body moving to keep it strong and healthy."

"Okay."

"And I've been thinking about going to this Zumba class at the gym; they play fun music and exercise at the same time. I'd love it if you came along. I'm sure you'd be better at it than me!"

"Oh, I definitely would be. But I'm not sure."

"Okay, well how about you try it once and if you don't like it, then you don't have to come again. How does that sound?"

"Okay, that's fine."

Remember, we don't want to overload our child right away, and we want to make sure they are interested in what we are offering; otherwise, they won't want to follow through—remember **the importance of play**.

If (or should we say "when") you are met with resistance, don't let it pull you into your Downstairs Brain. We are asking our child to make some big changes, and it will take time and patience for everyone involved.

DIGITAL DISRUPTION
SCENARIO #9

"My 9-year-old wants to play video games that I think are too violent for his age and he gets mad when I tell him he can't. Today I found him playing a game he wasn't supposed to be playing. I know I will either end up yelling at him, or just giving in because it's easier than dealing with his anger."

DTAP® Concepts to Think About		
Downstairs Brain	**Midbrain**	**Upstairs Brain**
Structure and Nurture Window of Tolerance	Verbal Responses Sitting in the Yuck	Natural and Logical Consequences

In the moment, here are some things you need to remember before you respond:

School-age children have to learn to navigate social norms, etiquette, and pressure, and many kids have access to all kinds of games and digital content. Remember, it can sometimes be really hard to be the kid who doesn't have the thing that everyone else seems to have: You will need to **sit in the yuck** with your child as you help them understand this reality.

Despite being almost double digits, a 9-year-old's brain isn't fully developed. This can make it hard for them to understand, and remember, why you have certain expectations and boundaries. You will need to balance **structure** with **nurture** and keep your **verbal responses** clear and concise to help them develop these skills.

One other thing. Our kids are still working on their ability to manage, process, and express their emotions, so it's not uncommon for them to have mini meltdowns from time to time. It's important we stay out of our Downstairs Brain (by paying attention to our **window of tolerance**) when this happens; otherwise, that mini meltdown might turn into a massive one, and not just for your kid.

Here's what to say to manage this *in the moment*:

Parent: Hey, kiddo, I need to chat with you. I'm going to set a timer for three minutes. When the timer goes off you need to pause so we can chat.

Child: Okay.

"Our timer is going off, so I need you to pause the game. Do you remember that we decided together that this isn't a good game to play?"

"Everyone else plays it! You're so unfair!"

"I know it feels really unfair right now."

"Why can't I play it?!"

"We can talk about it after we have both calmed down."

"I don't want to calm down!"

(Remain calm, and if it's safe to do so try to sit down and model taking deep breaths. Allow your child to pace and yell, as long as they are not hurting themselves, others, or property. This is **sitting in the yuck**.)

"I see you're feeling mad, that's okay. Do you need my help?"

"No!"

"Okay, I'll be here when you're ready."

(Continue to allow your child to move and make noise if they need to. This is their attempt to self-regulate and get back within their **window of tolerance**. When their movements subside and their breathing is settled, you can begin to address what just happened.) *See note on page 222.

"Phew, those were some big feelings you just had."

"I just want to be able to play the games I like."

"I understand. Can we talk about it some more?"

"Sure."

"Can you tell me what you like about that game?"

"I like that you have to shoot the bad guys and it's a fast game."

"It sounds like you like the action. The blood and the violence aren't very nice though."

"I don't care about that."

"I understand. Hey, you know you like science, right?

"Yeah."

"Well, did you know that even though you don't care about seeing all the blood going everywhere, your brain can actually start to think that it's normal? Sometimes it can also make you feel different feelings like worry and anger, even when you don't mean to."

"Really?"

"Yeah, so that's why we don't really like you playing those games all the time."

"I didn't know those games could make you do that."

"No, I didn't either until I looked it up. Want to look it up with me?"

"Sure."

Spend some time with your child looking up how certain games impact the brain—in an age-appropriate way. This is a great way to use technology to our advantage.

"So, I'm sure it feels like all your friends play that game, but can you understand now why we don't really like you playing it?"

(Here you are using **verbal responses** to validate your child's feelings.)

"Some of my friends play it. Yeah, I get it."

"And it makes us both upset when we have to remind you not to play it."

"Yeah, you got really mad."

"I did, you're right. I could have handled that better. I'm sorry. I wonder if there is a different game that still has action but not so much blood and gore that you could play?"

(Here you are balancing **structure** and **nurture**.)

"Maybe."

"Let's look together and see if we can find one that we both agree on."

"Okay."

*Please note: If behavior escalates or becomes physical, we encourage you to do the following:

- Remain calm—your regulation is vital, if you need to "tap out" with another adult to become regulated, that's okay. If no other adult is present, take deep breaths and count to 10.
- Stay in the same room or area as your child—we are looking for proximity (as this counts as **nurture**), but this is not necessarily the time for touch, as this can potentially escalate behavior further.
- If you're able to sit down, do so—this helps convey the message to your child's brain that you are not a threat (remember, when in Downstairs Brain, there is little access to the logical processes of the Upstairs Brain).
- Use as few words as possible—simple statements like "stop please," "I'm here," or "let's breathe together" are good options.

Now, here's how we *proactively* reinforce this:

Scenario 1: You find your child playing the game he wasn't supposed to be playing again.

Parent: Hey, kiddo, I need to chat with you. I'm going to set a timer for three minutes. When the timer goes off you need to pause so we can chat.

Child: Okay.

"Our timer is going off, so I need you to pause the game. Can you tell me why I might be feeling a bit frustrated?"

"Yeah, I was playing that game."

"Can you help me understand why you would still play the game after we agreed not to?"

"I just wanted to see what my friend was talking about."

"It must be hard when your friend is talking about something that you don't know about."

"Yeah."

"But you also know that you aren't supposed to be playing that game. Would it help if I took the game away?"

"No!"

"Okay, well we need to find a way to deal with this; otherwise, I will have to take it away. I wonder if I need to help remind you some more?"

"That would help."

"Okay, I can do that. Hopefully that helps, but if you can't manage that we will have to stop using the game until you can show me that you can follow the rules and we will have to play with our toys or books or go outside."

Scenario 2: Your child is doing great at following the rules.

Child: I'm going to play my video games!

Parent: Okay, hang on a second, kiddo, remember what we talked about the other day?

"Yeah, I have to play a different game."

"That's right, and I'm really proud of you for sticking to that rule so far. I'll keep reminding you so that you don't forget."

While kids' brains are still developing, it's normal that they might not be able to follow rules consistently after just one conversation. The most important thing you can do is stay calm, give reminders, and, if need be, follow through with **natural and logical consequences**.

DIGITAL DISRUPTION SCENARIO #10

"My 10-year-old comes home from school and gets to sit and play video games for two hours while I finish work, before dinner. By dinnertime she is so grumpy, and we always end up in an argument. I don't know what to do. She loves video games, so I thought she would be happy."

DTAP® Concepts to Think About		
Downstairs Brain	**Midbrain**	**Upstairs Brain**
Attunement Structure and Nurture Window of Tolerance	Verbal Responses Sitting in the Yuck	Problem-Solving and Abstract Thinking

In the moment, here are some things you need to remember before you respond:

Kids at any age need plenty of movement; however, this is something we don't understand until we are grown-ups! And unfortunately, many educational systems have reduced the amount of physical activity students now get, meaning that your child may well be sitting for most of their school day. So, by the time they are home they have movement needs that haven't been met. They are at the edge of their **window of tolerance**, but that video game is just so tempting, due to the large dopamine dose it delivers to the brain (that's the "feel good" chemical to you and me), so they choose that over anything else. We must use our **attunement** skills to help them recognize and understand that their body needs to move.

We will need to provide **structure** and **nurture** to help engage them in physical activity and be ready to use **verbal responses** to validate feelings of sadness or frustration relating to not being able to play their game.

If your child has become used to a particular routine, we have to acknowledge that it might be hard for her to immediately stop and do something else (remember that transitions can be hard for everyone, but especially kids). So instead of enforcing an immediate stop, we are going to give you an example of how to proactively implement gradual changes to help make the transition easier for everyone. You're probably going to have to **sit in the yuck** for a bit while she gets used to the new expectations.

However, just in case you've found yourself in the thick of it with a grumpy pre-teen . . .

Here's what to say to manage this *in the moment*:

Child: Why do I have to sit at the table to eat dinner?

Parent: Because it's important for us to eat as a family.

"I don't want to sit at the stupid table."

"I get that sitting at the table as a family can feel silly sometimes. It's important so we can catch up with each other."

(This is a great example of **verbal responses**.)

"Yeah, it's not fair that I have to sit at the table."

(Try to remember that it's not *really* about sitting at the table.)

"I hear that things feel unfair right now. Thank you for letting me know. Would you like to help me plate dinner up or would you like to sit at the table and wait?"

(Here you are **sitting in the yuck**, not changing things, just accepting them as they are.)

"Ugh." (Child sits down.)

"Great choice!"

Note: We are still not going to address the routine change yet. Everyone (including us) needs time to calm down, and our child definitely needs to eat before we can attempt **problem-solving and abstract thinking**. Remember, this doesn't mean she's "getting away with" a challenging behavior; it means we are choosing to handle it in a smarter, more effective way.

Now, here's how we *proactively* prepare for, and reinforce this:

In the morning, before school, or even better, on the weekend . . .

Parent: Hey, kiddo, when you come home from school today, after a snack, we are going to talk about our after-school routine.

Child: Okay.

After school . . .

"Hey, honey, now that you've had your snack, we need to talk about our routine after school. First, can you tell me a bit about your school day. Do you get to play and run around at all?"

"Well, we do PE on a Tuesday and Thursday."

"Okay, do you have recess?"

"Sometimes but not if it's rainy or cold."

"Ah, okay, that makes sense. And the rest of the time you're sitting at a desk working, right?"

"Yeah mostly."

"Hmmm, it doesn't sound like you get to burn much energy then, or move around a lot."

"No, not really."

"And what have we usually been doing when you get home from school?"

"Well, you work and I play my video games."

"Right. I know you enjoy your video games and it's a nice break from schoolwork. Can I tell you what I notice though?"

"Sure."

"Well, it seems like at dinnertime we both get a little cranky."

"Yeah, we do."

"I wonder if it's because we haven't had enough movement time during the day and maybe not enough outside time on nice days?"

"Maybe."

"I'd like to try something different to help us both be less cranky. I'm going to give you an idea and you can let me know what you think."

"Okay."

"So, how about when you come home and have had a snack, we do thirty minutes of outside time or dance time using a video on the TV, then you can have thirty minutes of video games, then another thirty minutes of outside or dance time, then video games until dinner is ready?"

"Can I play video games first?"

"I'm okay with that, but what happens if it's hard for you to stop when the timer goes off?"

"I could lose the second thirty minutes?"

"That seems reasonable. How about we try it tomorrow, and after dinner we can talk about how it went?"

"Okay!"

"I'm glad that we worked through this together. I don't want either of us to be cranky anymore."

We wish we could tell you that this conversation will fix everything right away, but it probably won't. What we can tell you is that with practice, reminders, and patience (and a little more **sitting in the yuck**), your child *will* get used to the changes. And everyone will be better for it.

TOP TECH TIPS FOR THE CHILDHOOD YEARS

1. Instead of cutting out technology completely, use it to your advantage. Find ways to incorporate it into activities such as exercise, cooking, academics, or organization, but keep it balanced with screen-free time.

2. Have patience and be prepared to practice. Cutting down on tech time will be a process and one which you will need to support your child through. Be okay with mistakes, and remember your ultimate goal is to teach, not punish.

3. Install parent/caregiver controls on any devices your child has access to. This is not an invasion of privacy; it's a way to keep your child safe.

4. Create dedicated tech-free spaces and times in your home, for example the dining table and the backyard.

5. Have set **structure** for screen time, including when and where screens go at bedtime.

QUICK SWAPS TO TRY

Instead of this...	Try this...
Setting all the screen-time rules	Create a screen-time contract together that everyone signs.
Asking your child to switch off a digital device immediately	Set simple **structure** like "after every thirty minutes of screen-time, participate in ten minutes of inside or outside physical activity.
Ignoring difficult conversations about digital disruption	Narrate your own screen use: "I am going to put my phone down now to play with you."

Simple actions, repeated consistently, make a real difference. And remember, you are still your child's parent—that means you are allowed to set **structure**, even if it's hard for your child to accept. Meet them with kindness, use your **verbal responses** to validate their feelings and help them by engaging them in other activities that remind them that there's fun to be had outside of the screen.

fridge-worthy facts

- Children don't need us as much as they used to, but they do still need us, especially when it comes to setting **structure** around technology use.
- Digital media can be used to enhance learning and **play** but shouldn't be the only source.
- Change won't happen overnight. Your child's brain (and body) needs time to adjust to new or revised expectations. Everyone gets grace.

Chapter 8

STRATEGIES TO MANAGE DIGITAL DISRUPTION IN THE EARLY TEEN YEARS (AGE 10–AGE 14)

"Growing Up Online Without Losing Connection at Home."

The young teen (or tween) years can be hard for everyone. It's a time when our child's world is growing (not to mention their bodies!), and they are desperately seeking more independence, while we adults are often left wishing that we could snuggle and protect them for just a little bit longer.

Hormones are rampant, which causes your tween to spend a lot of time in their Downstairs Brain (and if you aren't careful, you too), while they figure out all the ways their bodies and emotions are changing. It's important to remember that adolescent development is gradual and differs for every child, meaning that what might be right for one child when it comes to digital media could be different for the next. This is especially true for social media use.

Vogels and colleagues (2022) identified three main things you want to think about when it comes to young teens and digital media use:

1. The amount of time they spend on digital media.
2. The type of content they consume or are exposed to on digital media.
3. The degree to which their time, consumption, and exposure might be disrupting activities that are essential for overall health.

The DTAP® interventions you will see in this chapter will help you teach your child how to balance their use of digital media depending on their level of maturity. Some young teens may need more **structure** than others. Some will only need prompts or gentle reminders about staying regulated or thinking through some situations. For others, or in more complicated situations, you may need to be intentional with your **verbal responses** and be prepared to **sit in the yuck** as they work to balance their lives, make decisions, and experience extreme emotions.

Because each young teen is different, it is important to have open conversations about screen time and discuss limits based on their specific needs and activities. Some are still not ready to manage the responsibilities that come with unlimited access to digital media.

Finally, because a young teen's brain is still developing, we will need to offer continued support with **independent regulation**, **problem-solving** and **abstract thinking**. And it goes without saying, but you're likely going to have to enforce some **natural and logical consequences** too.

Stay strong. We've been in these trenches, and there's light at the end of the tunnel, we promise!

DIGITAL DISRUPTION
SCENARIO #11

"My 11-year-old has started to self-harm. I am concerned that some disturbing content he saw online is what triggered his cutting. I know he is hurting and he feels terrible about it."

DTAP® Concepts to Think About		
Downstairs Brain	**Midbrain**	**Upstairs Brain**
Attunement Structure and Nurture Window of Tolerance	Verbal Responses Sitting in the Yuck	Problem-Solving and Abstract Thinking

In the moment, here are some things you need to remember before you respond:

Self-harm behaviors can be very scary for parents, caregivers, and kids. You don't have to have all of the answers right away. It's okay to give yourself time to process your feelings and think about what you want to say, but it's important that you don't avoid talking about it—your child needs you to be strong for them emotionally. It is going to be uncomfortable and will require you to **sit in the yuck** with them and with your own feelings.

Because this will be a difficult conversation for everyone, it's important to remember your child's **window of tolerance** and be **attuned** to behaviors that indicate they are nearing the edge of their **window**. If you do spot these behaviors, it's okay to take a break from the conversation and return to it when everyone is ready again.

Try to remember that young teens can experience extreme emotions and tend to see things as all good or all bad. They might need help understanding that they are not damaged or flawed. Your use of **verbal responses** will help them experience and process their emotions.

There might be shame or embarrassment related to the content your child has accessed. While your initial reaction might be to increase **structure** (out of concern for safety), it's important to balance this with **nurture**. Assure them you do not feel differently about them, listen without judgment, and let them know you understand their feelings.

Note: It is also okay to *not* understand and it's more than okay to say, *"This is hard for me to understand right now but I'm not going anywhere and we will get through it together."*

Additionally: Please refer to page 307 for details of professional resources relating to self-harm and suicidal behaviors. It's ok to seek outside support when helping your child **problem-solve**.

Here's what to say to manage this *in the moment*:

Parent: I noticed that you have some cuts on your arms. Would it be okay if we talked about that?

Child: I guess.

"Can I get you a drink or a snack before we do?"

(Being **attuned** to needs prior to a difficult conversation can be helpful, it's also a way to provide **nurture** and communicates to our child, *"I'm here to take care of you."*)

"Sure, can I have a soda?"

"That's fine. Can you tell me about the cuts on your arms?"

"I mean, I don't know, I just did it."

"Sometimes we do things when we are young and don't really know why. I'd like to see if I can help you figure it out."

"Okay."

"I've noticed that you seem to be really hurting right now, feeling quite sad. I wonder if that has anything to do with it?"

(Using the **verbal response** of wondering is a gentle way to explore big feelings.)

"It started just after I was searching for ideas for a Halloween costume. I had pop-ups of other kids showing how they do it and saying it made them feel much better. I guess that is why I started doing it too."

"Oh gosh, it sounds like you thought it was a good solution, but hurting yourself is very dangerous."

"I am embarrassed that I even looked at those videos. Are you mad at me?"

"I am not mad at you; I am so pleased you are telling me. I want to keep you safe."

(Child doesn't respond and that's okay—this is part of **sitting in the yuck**.)

"I know this is a hard thing to talk about; maybe it's even a bit scary."

"I don't like feeling this way."

"I bet it doesn't feel nice at all. Let's work together to find other ways to help you feel better."

"Can I just drink my soda?"

(This is your child's way of saying, *"I'm reaching the edge of my **window of tolerance**, and I need to take a break from this conversation."* It's okay not to force the issue and come back to it later in the day or the next day.)

"Man, I tried to jump into fix-it-mom mode, didn't I? Sorry about that. I am so glad we had this conversation. Let's talk some more tomorrow when you're ready."

Now, here's how we *proactively* reinforce or follow up on this:

Parent: Hey, kiddo, come have a seat next to me. We need to finish the conversation we started yesterday.

(Encourage your child to sit next to you, not in front of you. This arrangement is less confrontational for kids. Even better, if you can have the conversation during a calm car ride, the rhythm and movement of the vehicle can be soothing for everyone.)

Child: Okay.

"I'm so glad you told me about the things you saw online and the cutting you were doing. That must have been pretty scary for you."

"Yeah, I was worried about how you would react."

"It was a bit scary for me too. But that's okay. I'm here for you and we can get through this tricky spot together. You said yesterday that you don't like feeling this way. Is that true still?"

(Have courage to **sit in the yuck** with your child during this difficult conversation. It's okay to admit when things are a bit scary for you too, but be sure to balance that with reassurance that you're there to support them through the tough stuff—remember, object constancy (page 39 can tell you more about this if you need a reminder.)

"Yeah."

"Okay, I have some ideas for helping you to feel better. Can I share them with you? And then you can tell me what you think."

"Okay"

"Would it be helpful if you had someone else other than me or your dad to talk to about how you feel? It's okay if you say yes; sometimes it's good to talk to someone different."

"Like a counselor?"

"Yeah, that's what I was thinking."

"Okay, but not at school. I don't want people at school to know."

"That's fine, we can look together and find someone you are happy with."

"Okay."

"And I'd like to know how I can help you when you feel sad. Is there anything that helps you to feel better other than cutting on your arm?"

"I like playing on my tablet, and when we play board games together." (Gently help your child think about **problem-solving** while not distracting from the difficult conversation).

"Okay, that sounds good, I like board games too. You and I are going to take a look at the settings on your tablet. We might have to make some changes so that those kinds of videos don't come up again. Even though it made you feel a bit better, it's not a safe option, can you understand that?"

"Yeah."

"Okay. One other way I would like to help is by doing a check-in with you each morning and evening. It can be something simple like you give me a thumbs up if you feel good, down if you feel sad, and in the middle if you aren't sure. How does that sound?"

"Yeah that's fine."

"Tell you what, we will try it, and after a few days we can decide if it's helping or not. Is that okay?"

"Yeah."

Having this conversation with your child helps them know that they are not alone, and that you are able to support them even when things get scary or uncomfortable. This is one of the roots of secure attachment—**attunement**, **verbal responses**, and **sitting in the yuck** are all concepts that can help you do this.

Also, it's important to consider what support *you* might need in order to be a place of emotional safety for your child—discovering self-harm behaviors can be frightening and overwhelming, and can push us into our own Downstairs Brain. Confiding in a trusted family member or friend, or a professional therapist allows space for you to process and manage your own strong feelings, so that you can help your child do the same.

DIGITAL DISRUPTION
SCENARIO #12

"My 12-year-old child spends every weekend glued to social media and then gives me a play-by-play of friends who are at parties or sleepovers to which she was not invited. She used to have fun with us at home, but now she is obsessed with seeing where and with whom she was not included. She spends more time crying on the weekend than not."

DTAP® Concepts to Think About		
Downstairs Brain	**Midbrain**	**Upstairs Brain**
Structure and Nurture Window of Tolerance	Verbal Responses Sitting in the Yuck	Problem-Solving and Abstract Thinking

In the moment, here are some things you need to remember before you respond:

For young teens, friendships are their whole world; in fact, we all have a primal need to belong and to be connected to others. It can be excruciating for our kids growing up in the digital age because many teens document their lives on social media, and when they see everything their friends are doing, fear of missing out (FOMO) becomes real and unavoidable.

We need to offer our young teen empathy, hold space for them to vent, and listen to their feelings without offering guidance—really **sitting in the yuck** with them. You might share stories of when you felt similarly, as this helps teens feel like they are not alone. This will be hard, and you will need to be aware of their **window of tolerance** for big emotions.

We will need to use a balance of **structure** and **nurture** to help set expectations around social media use, while using **verbal responses** to help our child know that we understand their feelings. You might explain that things posted online aren't always "real life" and that people only share what they want others to see—this helps them with their **abstract thinking** skills.

To help support **problem-solving**, invite your teen to watch a movie or bake cookies with you or another family member, encourage them to make plans with another friend, or engage in one of their hobbies on the weekends. Perhaps even take a trip somewhere fun where they can take their own pictures to post to social media if that feels appropriate for your family.

Here's what to say to manage this *in the moment*:

Parent: I'm sorry you are so upset right now.

Child cries.

"I'm going to come and sit by you."

(Close proximity, and even a hug if it feels appropriate, will provide **nurture** to your child.)

"Want to tell me what's going on?"

"All of my friends are having sleepovers without me."

"I'm sorry to hear that. I wonder how that makes you feel?"

(Our **verbal responses** will begin to allow our child to explore her feelings.)

"I hate them!"

"That's a big feeling! I'm glad you shared that with me. It sounds like you're angry."

"I am."

"I notice that you are crying also. I wonder if you are sad too?"

"I guess so."

"I think I might feel both of those things too if I saw some friends spending time together without me."

(Be okay with **sitting in the yuck**, and continue to provide **nurture** with gentle touch and close proximity.)

"It can be really hard when we see people doing fun things. It feels like we might be missing out."

Child does not answer. This is okay.

"Hey, why don't I make us some hot chocolate and we can talk about this a bit more?"

(Being aware of your child's **window of tolerance** is important. It's okay to offer them a break when things get overwhelming.)

"Sure."

Now, here's how we *proactively* reinforce or follow up on this:

Parent: I've noticed that you seem to be upset more on the weekends lately. Can you tell me about that?

Child: It's because I see all my friends hanging out together or doing fun things. I never get invited, and we never do anything fun!

"I'm sorry that you feel like you are missing out on the weekends and that it feels like we don't do anything fun. Can I tell you something?"

"Sure."

"Sometimes it seems like all you want to do is be on your phone most days. I feel like you don't want to be around us."

"I do, but my friends are all online."

"Okay, well that makes sense to me. I think I hear you saying that you feel left out, but did you know, often, people only post things online that they want others to see."

"What do you mean?"

"Well, we went to the mall last weekend, right? And you took a few fun pictures of your new outfits and that fun drink we got from the coffee shop."

"Yeah! It was fun! I put those pictures on my social media account and people liked them."

"Right, but what did we do for the rest of the weekend?"

"Ugh, we had to pull all the weeds in the yard, do laundry, and get groceries."

"That's right. It wasn't very fun was it?"

"Not really!"

"So, I want you to think about the things that your friends post online and your fear of missing out. Yes, sometimes you might miss a few things, but I promise you that the rest of the time they are just doing normal things like us."

(Now that our child has worked through her big emotions, she is ready to engage in **problem-solving** and **abstract thinking**.)

"I hadn't thought about it like that."

"That's okay, that's why I'm here. To help you understand these things. I wonder if it would be helpful for us to do a check-in on the weekends to see how you are feeling and if you need my help?"

"That might be helpful."

"Okay, I can do that. Also, I wonder if it might help if we invite a friend over for a sleepover here from time to time? What do you think?"

"That would be fun! Thanks!"

This is likely a conversation that will have to be revisited time and again as you help your young teen learn to navigate the world of social media and peer groups. And we can tell you from personal experience that there will be more tears and frustration (for everyone!) However, if you remain consistent in your approach and open with communication, your relationship will stay strong, despite the minor ruptures that happen. Don't forget, deep down our kids still need us, even when they seem to be pushing us away.

DIGITAL DISRUPTION SCENARIO #13

"My 13-year-old daughter complains about being tired and falling asleep in class. I think she is staying up too late, and tonight she said she was going to Facetime her friends to work on homework together. I know if I challenge her it will turn into an argument, but she needs her sleep."

DTAP® Concepts to Think About		
Downstairs Brain	**Midbrain**	**Upstairs Brain**
Attunement Structure and Nurture	The Importance of Play Verbal Responses	Problem-Solving and Abstract Thinking Independent Regulation Natural and Logical Consequences

In the moment, here are some things you need to remember before you respond:

The life of a young teen (even before the invention of cell phones and tablets) revolves around friends, and more often than not, their intentions as a group are not to cause trouble, but sometimes their behaviors are problematic, hurtful, or skewed because of that underdeveloped Upstairs Brain. Meaning, they want to do good but sometimes they mess up. Use **attunement** and **verbal responses** to let your teen know that you understand this.

Remember that your young teen is still learning how to **independently regulate**, and sometimes we might have to let them experience **natural and logical consequences** to help them understand the importance of paying attention to their needs. That doesn't mean we can't support them by setting **structure** with a dose of **nurture** for balance.

If it's late at night, it's not the time to get into it with your child. Tiredness leads to Downstairs Brain behavior for everyone, and we may all end up saying things we don't mean (much like we do when we are hangry!) Upstairs Brain skills such as **problem-solving** and **abstract thinking** are less likely to be accessible at this time.

When you do address the issue, try to remember the **importance of play**; there might be an opportunity to bring humor to the situation to keep everyone out of their Downstairs Brains.

Here's what to say to manage this *in the moment*:

Teen: I'm getting on Facetime with Sarah and Ali at eight thirty to work on our group project.

Parent: That seems kind of late. How much more of the project do you have left to do?

"Not much."

"Okay, and when is the project due?"

"Not until next week."

"Okay, I know that it's important to you to get your schoolwork done. What time do you think you'll be finished?"

(This is a great **verbal response**, which diffuses a situation that could quickly escalate. Remember that your teen is likely expecting you to challenge them.)

"I don't know, maybe ten thirty."

"Okay. Would tomorrow after school be a good time to talk about this some more? I just want to make sure you are still getting enough good sleep, especially as you mentioned falling asleep in class earlier this week."

(This approach is called choosing your battles. We say this with kindness(!); we have been there and felt the frustration, but remember that this is not the time for **problem-solving** and **abstract thinking**.)

"Sure."

"Okay, I'll say good night when I go to bed. Try not to stay up too much later than ten thirty p.m."

(This small dose of **structure** is important, even if it's not the preferred option right now.)

Now, here's how we *proactively* reinforce this . . .

Parent: Let's talk about last night and some of the troubles you've had falling asleep in class.

Teen: Okay.

"I wonder if you have been staying up too late on school nights?'

(This **verbal response** gently shares your perspective but also communicates that you want your child's input too.)

"Well, I'm just doing homework with my friends on Facetime."

"I understand that, but it sounds like you are not feeling very well-rested, and I know what you girls are like when you get together; you could talk someone's ear off!"

(Remember, **play** and humor can help diffuse a tense situation because it signals to your teen's brain that you are not a threat.)

"Yeah, maybe we spend more time discussing makeup and boys more than we should."

"Oh yeah, I was a teen at one time too, although we only had one phone in the house and it was not to be used after nine p.m.! So, you need to get some good sleep; otherwise, you'll be falling asleep in class. What do you think we could do to make sure that happens?"

(Here you are helping your teen recognize **natural and logical consequences** and use **abstract thinking** skills.)

"Well, it's hard to stop everyone talking when they get going, and I'm embarrassed to tell them that I fall asleep in class and have to go to bed early."

"I understand. How about if I see that you're still on Facetime at nine thirty, I'll come in and tell you it's time to get off, and that

way, you can blame me and say, 'Ugh, Mom says I need to get off.' Would that work?"

(Using a balance of **structure and nurture**, this **verbal response** demonstrates empathy but also helps your teen with their **independent regulation** skills.)

"Yeah, we could try that."

We'd be lying if we said that it's as easy as the example above, we know that it isn't. If our own personal experience is anything to go by, we can definitively say that it's easier said than done! Young teens seem to love to test limits (although in all honesty, they usually don't think about it to that extent, they just *do*. With that in mind, we wanted to walk through how we would handle it when (not if) our teen tests the **structure** we have set.

Parent: Hey kiddo, once you've had a snack we need to talk about what happened last night after I told you it was time to get off your phone.

Teen: Okay.

"So, just like we planned, I came up at 9:30 p.m. and told you it was time to put the phone up. I heard you say goodnight to your friends, which is great. But, when I went to get a drink of water at 10:15 p.m., I heard you on your phone again. Can you tell me what that was about?"

"I just put a quick video on before I went to sleep!"

"Hmm, okay. Well, I happened to peek out of my room again at 10:40 and I could still hear you talking. Can you tell me about that or should I have a guess?"

"Okay, I was on a Facetime call."

"Thank you for being honest with me. It seems like our strategy for ending your calls at an appropriate time so that you can get a good night's sleep isn't working. Unfortunately, this means that you will need to give me your phone at 9:30 p.m. instead."

(This is an appropriate **natural and logical consequence** based on our child's actions.)

"Wait, what?! That's not fair!"

"I understand that you feel upset about this. I also know that you need to be able to focus at school and not fall asleep in class."

"I hate you!"

"It's okay if you're mad at me right now. Part of my job is to keep you safe and healthy."

(This is **sitting in the yuck**—accepting our child's feelings but maintaining our [**natural and logical**] consequence.)

"Leave me alone!"

"Sure, I'll check back with you later."

Sometimes we all just need some time to calm down. This is especially helpful in avoiding bigger, more hurtful arguments, plus it gives our young teen the opportunity to get out of their

Downstairs Brain. And remember, just because you are giving space and time, it doesn't mean that the expectation has to change. Here's how we would follow up with our young teen when they are calm again:

Parent: "Hey honey, we need to have a chat about what happened earlier. Would you like to talk now or after dinner?"

(Allowing your child to have a choice is a way of reminding them that you value what they have to say and can make it easier to have tough conversations.)

Teen: "We can talk now."

"Okay, you were pretty upset earlier. I wonder what was making you mad?"

"You were. You said I had to give my phone to you every night at 9:30 p.m. No one else has to do that."

"That's right I did. Why do you think I said that?"

"Because I made a call when I shouldn't have."

"That's right. We had an agreement."

"But it's so unfair, I don't want to give my phone up."

"I get that it's hard for you, especially as all your friends have phones too. It seems like it's too much for me to expect you to handle that all by yourself, which is why we are going to start by putting the phone in my room every night at 9:30 p.m. Perhaps we can practice a

curfew on the weekends and if you get good at sticking to that then we can try again during the week."

This is a situation that many parents and caregivers find themselves in with their kids. To be honest, it's not a new one—even before the advancements of technology kids would stay up too late. However, it's still important to address this, as sleep is crucial for our young teen's brain and body development. We have found that getting our child's input can be really helpful in this particular scenario, and being prepared to compromise can go a long way. It's also okay to increase **structure** though if your child is not able to manage more freedom yet.

Remember, do what works for you and your family.

DIGITAL DISRUPTION SCENARIO #14

"My 14-year-old daughter hangs out on an online platform and claims she is meeting all kinds of nice gamers from around the country, but we do not know these people, and I read something about groups and predators known to target children for sexual exploitation using this platform to communicate with children. She's also spending more time online than with real-life friends."

DTAP® Concepts to Think About		
Downstairs Brain	**Midbrain**	**Upstairs Brain**
Window of Tolerance Structure and Nurture	Verbal Responses Sitting in the Yuck	Problem-Solving and Abstract Thinking

In the moment, here are some things you need to remember before you respond:

Your young teen is still developing their **problem-solving** and **abstract thinking** skills. They may not understand the dangers of online predators, and talking to them about this topic is important, even if it makes you uncomfortable—you should be prepared to **sit in the yuck** with your teen.

Young teens are most vulnerable because they are curious and want to be accepted. Be especially cautious if your child struggles with in-person friendships and is desperate for connection and belonging. Normalize the positive feelings that might come with online relationships: chats that express a teen's importance and someone's desire to pay attention to them. Sometimes, teens believe they are in love with someone online, making them more likely to agree to something risky, like sending inappropriate pictures. We should use **verbal responses** to explore this with them gently and with empathy.

Pay attention to your own **window of tolerance** and whether you might be pushed into your Downstairs Brain because of the sensitivity of the conversation. It's always okay to take a break and come back to it when you are feeling regulated and in control.

You will likely have to enforce **structure** around your child's use of the digital platform—and with her desire to be independent it's likely this won't be received well. It's important to remain firm but remember to increase **nurture** in order to make this transition easier.

Here's what to say to manage this *in the moment*:

Parent: Hey, honey, I'd like to talk to you about the people you have been chatting with online. Is there a good time we could do that?

Teen: Um, okay. We can talk now.

(Find a place to sit next to each other, so that you aren't directly facing your teen. Conversations are less intimidating this way. Or you could even suggest driving somewhere to take a walk and talk on the ride.)

"So, I remember you saying that you've met lots of nice people online. Well, you know me, I'm old and I don't really know much about this stuff! How do you know they are nice people?"

"Well, they are friendly and kind, and they say nice things to me."

"Okay, that's good. And where in the world are they?"

"Oh, all over!"

"Are there any that you chat to a lot, or that you've gotten to know well?"

"Yeah, a couple of girls from the US and a few boys."

"Oh, that's great! I noticed you got a big smile on your face when you mentioned the boys. Tell me about that!"

"Mom!"

"It's okay, you can tell me!"

"Okay, well there's one boy, he actually lives in the next state!"

"Oh, really, how do you know that? Did you tell him where you live?"

"Only what state I am in. We've been chatting a bit; he's really funny."

"Okay, that's good. I did notice that you spend a lot of time online. Are you talking with him all the time?"

"I guess we do talk a lot."

"And what do you talk about?"

"All kinds of things."

"So, do you know what he looks like and how old he is?"

"Well, he doesn't have a picture, but he's told me what he looks like and he said he's 15."

"Okay, and you have a picture of you online, right?"

"Yes, just a small one."

"Has he asked to see any more pictures of you."

"No, but he did tell me I was pretty."

"Well, that's nice, but you should definitely be wary of someone flattering you online."

"He said he'd love to meet me one day, as we live close to each other."

"I'm curious, how do you know that he is who he says he is?"

(Using **verbal responses**, such as wondering with curiosity, can help your young teen develop their **abstract thinking** skills.)

"What do you mean?"

"Well, how do you know that he is actually a 15-year-old boy? If you haven't seen him in real life, then is it possible that he is a grown man pretending to be a boy?"

"Well, I have heard his voice, and he doesn't sound like a man."

"Okay, well that's something. But I want you to be careful. I don't want you sending any pictures to him until you have seen his face. Perhaps you could ask him to video call you sometime while I am around so we can see him?"

(Here you are suggesting a way to **problem-solve**, while not shaming your young teen.)

"I guess so."

"I know it seems silly, but there are people out there who pretend to be people they aren't so that they can get young girls interested in them. They even add your friends online too. They make you like them and then they ask you to send inappropriate pictures to them. They then threaten to post the pictures online if you don't send them money. So we need to be really cautious when talking to people on the internet."

"Oh wow, okay."

"I know this is a lot to think about. I'm here if you want to talk more about it."

(This could be a good time to take a break from the conversation—remembering your child's **window of tolerance**. You have given her a lot to think about, and she might be feeling overwhelmed.)

Now, here's how we *proactively* reinforce this:

Parent: Hi, honey, I wondered how things were going online, particularly with the boy you were telling me about?

Teen: Well, I asked him if he wanted to video chat so that we could see each other, and he ghosted me.

"What does that mean, ghosted?"

"It means he just stopped responding to me. I haven't heard anything from him in two weeks. I miss him."

(Inside, you might be feeling very relieved to hear this; however, we have to remember that for our teen this was a big deal. This is an opportunity to strengthen your relationship by **sitting in the yuck** with them and allowing them to share with you.)

"I'm sorry, honey. It sounds like you enjoyed talking with him."

"I did."

(This is a good time to provide **nurture**; a hug or close proximity is a good idea.)

"Why do you think he ghosted you?"

"I don't know, maybe you were right? Maybe he wasn't who he said he was and couldn't video chat because of that?"

"I guess that's possible. That's the problem with the internet; you have to be so careful about what you believe! I think this is a good lesson for us. We always need to make certain who we are talking to and not share any personal information such as where we live, or go to school, and definitely don't share any private pictures."

"Right."

"I wonder if you might like to hang out with some friends from school? I could take you to the mall or to a movie?"

(Once we have helped to process big emotions, we can begin to introduce **problem-solving**.)

"I've been wanting to go to that new skating rink. Could we do that?"

"That's a great idea! I think it will be helpful for you to spend some time with friends in the real world. That way you know exactly who they are!"

The key to supporting our young teens as they navigate the digital world is to be open and honest with them (in an age-appropriate way). When we demonstrate that we are comfortable having these conversations, we communicate to our child: *"You can talk to me about this stuff and I will love you just the same."* That's attachment 101.

It won't always be easy, and we might have to set aside our ego as we realize that we don't always understand the world our kids live in nowadays—yeah, it's a little embarrassing but we promise you aren't the first parent to have to do it! And wouldn't we prefer that to our child getting into dangerous situations online?

TOP TECH TIPS FOR THE EARLY TEEN YEARS

1. Don't be afraid to set limits with your child regarding screen use and social media use. While this is a time for

independence, it is still a time of high risk—our kids still need us to parent them.

2. Try to understand that our young teens live in a different world than the one most of us grew up in, and they are trying to navigate the digital era, just like us. It's especially important to help them carve out screen-free time to allow their brain to take a break from all of the digital distractions and social overload.
3. There will be times when you must have difficult conversations. You may even have to address certain topics earlier than you would have liked to! Young teens are now exposed—often unintentionally—to media content many adults wouldn't choose for them.
4. Young teens face real digital risks, and it's our job to help them realize and remember it.
5. Try scheduling a weekly tech-check in with your young teen—no lectures, just curiosity.

QUICK SWAPS TO TRY

Instead of this...	Try this...
Setting all the screen-time rules	Create a screen-time contract together that everyone signs.
Not addressing online activity because it makes you uncomfortable	Break down difficult conversations—take a break when you or your child needs it. And don't be afraid to share some of your discomfort with your child.
Getting in power struggles and arguments about screens, tech, and internet use	Find a time to discuss issues when everyone is calm and has access to their Upstairs Brain (even you). "In the moment" is usually not the time to come up with solutions.

A little bit of connection goes a long way especially when our teens are trying their best to distance themselves from us! Don't be too hard on yourself when things get tricky, or messy—what's most important is that your child feels emotionally safe enough to come to you, not how perfect you are.

Chapter 9

STRATEGIES TO MANAGE DIGITAL DISRUPTION IN THE LATE TEEN YEARS (AGE 14–AGE 18)

"Balancing Freedom, Trust, and Tech"

The older teen years might be a little less challenging than the earlier ones (phew), but they still have their inherent difficulties. The need for independence continues to increase, as do academic responsibilities and extra-curricular commitments. Older teens are in a constant battle of balance, all while trying to figure out who they are in the greater scheme of life.

While their brains are much better developed at this point, we have to remember that full maturity doesn't happen for at least another ten years, and so our small adults still very much need us big adults to guide, support, and (occasionally) correct them.

Older teens cannot consistently apply **abstract thinking** because strong emotions continue to drive decisions where impulses come into play. This is often the source of arguments they experience with parents. They may struggle with the ability to think ahead to the potential consequences of their actions. They will need a trusted adult to help guide them through the critical thinking necessary to **problem-solve** and make choices surrounding digital media. Depending on the outcomes of their decisions, this is a time where it can be helpful to allow **natural and logical consequences** to play out.

Older teens are very concerned about their friends and what others think about them, especially their appearance. Peer pressure peaks during this time. Digital media can make this much more complex. As parents and caregivers, we need to be ready to support our kids with **nurture** as they learn to manage feelings that come from social media use.

We need to be prepared to **sit in the yuck** with our kids, especially if they find themselves caught up in challenges associated with digital media, including gambling, pornography, sexualized behaviors (such as sending explicit pictures to others), and accessing other inappropriate content. It's also possible that we may still need to set appropriate **structure** and limits for technology, in order to help our older teens make the best decisions.

The DTAP® interventions within this chapter will help you support older teens who are striving for increased independence but are still developing good decision-making skills.

DIGITAL DISRUPTION SCENARIO #15

"My 15-year-old daughter is constantly on social media apps, trying to recreate looks she sees online. Then, she posts photos of herself and starts counting likes and how many followers repost her content. She is never happy with the numbers and blames it on her appearance. She complains about her body size and shape. Lately, she has been restricting her food intake, and I have noticed that she exercises even after she goes to basketball practice. I'm worried about her."

DTAP® Concepts to Think About		
Downstairs Brain	**Midbrain**	**Upstairs Brain**
Attunement Window of Tolerance Structure and Nurture	Verbal Responses Sitting in the Yuck	Problem-Solving and Abstract Thinking

In the moment, here are some things you need to remember before you respond:

Your older teen's life likely revolves around social media. She is invested highly in preserving her online status making her especially vulnerable to peer feedback and social comparisons. We will have to use our **verbal responses** to create emotional safety before we try to address this situation. We have to help her know that we understand (or are trying to understand) that this is the world she lives in right now.

The risk of your teen developing an eating disorder is very real. In fact, some researchers have shown a correlation between an increase in eating disorders and body image problems and the emphasis on physical attractiveness online (Suhag and Rauniyar 2024). This is going to be a tough conversation for both of you—it will require **sitting in the yuck** and providing **nurture**.

You will have to be **attuned** to your teen's **window of tolerance**: The more her feelings get overwhelming, the harder it will be for her to stay inside of her **window**. Increasing **nurture** during this time may help, but also remember, it's okay to take a break and return to the conversation when everyone is regulated.

Your older teen lives in a digital world that is full of unrealistic ideals with trends and filters in the driving seat—combine this with her underdeveloped Upstairs Brain and you've got a recipe for a body-image disaster. We will need to spend time educating our teen about how photos online are often edited and that models and pictures are not usually representative of an average human's body. Reinforce that filters and preset edits are used to enhance or change a photo or video. You might also consider discussing the difference between health and "thinness."

Remember, **abstract thinking** is not fully developed, even in older teens; they lack perspective and cannot always think carefully about the media they consume unless it is presented to them. Encourage your teen to unfollow toxic accounts and instead follow accounts with diverse body types—if this is a struggle, we may need to set **structure** around their use of these platforms.

Here's what to say to manage this *in the moment*:

Teen: Why can't my body just cooperate with me? I am exercising and dieting, and I still cannot fit into any clothes I want to wear.

Parent: Honey, I think you have a lot of cute clothes, can you help me to understand what you mean?

"These girls post pictures of themselves, and the clothes look perfect on them; they are all so confident and happy. I want to look like them, and it is not fair."

"It sounds like you are frustrated and not feeling good about yourself. I am so sorry you are feeling this way. I understand that you care about how you look."

"You don't understand! I hate my body!"

"Ugh, I'm sorry, honey, I'm sure that doesn't feel good."

(If possible, provide **nurture** through close proximity or gentle touch. Be okay with **sitting in the yuck** of this big feeling, don't try to fix things yet.)

"You have a lot on your mind right now. How about we get some tea and swing on the porch and maybe we can chat about this?"

(You are recognizing that your teen has reached their **window of tolerance** with her overwhelming emotions. Suggest a break incorporating something she likes.)

"Ugh, fine."

Now, here's how we *proactively* reinforce this:

Parent: You know what, honey? I've been thinking, and you're right; I don't understand fully. The world you live in now is so different to the one I grew up in. We didn't have social media or all the platforms, and really the only people we compared ourselves to were the people we went to school with. Can you help me to understand?

(Using our **verbal responses** we are honoring our teen's experience. Spend some time just listening to your teen, again, without fixing.)

"I wonder, have you considered who you are comparing yourself to?"

(We can start bringing **problem-solving and abstract thinking** into the conversation if our teen is ready. This is done gently using our **verbal responses**.)

Teen: What do you mean? All these perfect girls!

"Do you think the likelihood that the photos you view are edited or using filters is extremely high?"

"I do not know, maybe. Are you just saying that because you are my mom?"

"No, I would say the same thing to anyone—the models and influencers on social media appear that way because often their images and videos are altered. There is research available that gives us evidence that this is the truth."

"Well, it still makes it hard."

"I understand that. There are a lot of external pressures for us to look a certain way, but they are not realistic, and perfection does not exist. My concern is that you will never feel good about yourself if you are always trying to look like someone or something that isn't real."

"Yeah, I have been having a hard time with that."

"I am so glad you shared that with me and I am sorry that it's been hard for you. Would it be okay for us to explore some ways I might help you?"

"Okay."

"Well, how about we take a look sometime at how pictures of people are edited? I'm sure there's plenty of videos online we can use."

"Yeah that could be helpful."

"I was wondering too if it would be helpful for me to check in with you regularly about how you are feeling? I want to be honest with you. I am worried about how much you are exercising and monitoring what you eat. I want to make sure you know I am here for you."

"Okay."

"I know it's not fun, having your mom check in, but making sure you are healthy and safe is important."

"Yeah, I know."

"And honey, if I'm worried about you, we might have to look into monitoring your accounts. I want you to be independent, but I also want to make sure you are seeing real, healthy content."

(Don't be afraid to set **structure** if you need to. Remember, our older teens still need us.)

"I get it."

We know this conversation might not go as smoothly as we've laid it out here, but our goal is to give you a good starting point for connecting with your teen about sensitive topics. The best way you can help is by not shying away from it, but instead, leaning into the discomfort. You are still your child's psychological home base, even when they are 15!

Note to parents: While eating disorders and body dysmorphia is more prevalent in females, these issues can also occur in males. We encourage the same approaches whatever the gender of your child.

DIGITAL DISRUPTION SCENARIO #16

"Help! Someone saw my 16-year-old son at a party and posted a photograph of him on social media passed out from drinking alcohol; he is angry and feels ashamed of himself."

DTAP® Concepts to Think About		
Downstairs Brain	**Midbrain**	**Upstairs Brain**
Window of Tolerance Structure and Nurture	Verbal Responses Sitting in the Yuck	Problem-Solving and Abstract Thinking Natural and Logical Consequences

In the moment, here are some things you need to remember before you respond:

While the news that your child not only engaged in alcohol use but that it was shared publicly may be upsetting and disappointing to hear, the way that we respond now (our **verbal responses**) may critically impact their decision to inform us and come to us in the future about other choices they make (or plan to make). You will need to begin by recognizing what part of the brain you are in and if you are nearing the edge of your **window of tolerance**.

If you are in your Downstairs Brain, it's not the time to address this. It is perfectly okay to say, *"I am really upset with you right now, and I need some time to process this."* Just make sure you return and reconnect with your child when you are regulated and have access to your Upstairs Brain.

For most parents, this conversation is going to take a lot of **sitting in the yuck** and that's okay. We get it. It's even okay for you to express this to your child.

It is important to remember that we were kids once, and we have all done things that we regret or should have done differently. In this scenario, our focus is going to be on the fact that nowadays our choices (both good and bad) are at risk of being critiqued by the whole world through digital media. We can do this by providing **nurture**, while also reinforcing our expectations (**structure**).

When everyone is calm, and out of their Downstairs Brain, we can start to consider **problem-solving** and **abstract thinking**—we might even find that the **natural and logical consequences** of this situation are enough of a deterrent to stop this behavior in the future.

***Please note: We in no way promote underage alcohol or substance use.** We are using this example to highlight the challenges teens and families face when unhealthy behavior choices are posted online.

Here's what to say to manage this *in the moment*:

Parent: I can't believe you did something so stupid! I trusted you to hang out with friends, and this is what you do?

Teen: I know, I'm sorry.

(Take some deep breaths. You are understandably in your Downstairs Brain, but staying there isn't going to help in the long run.)

"Look, I'm glad you told me this, but I'm really disappointed that you decided to drink alcohol, and I am really sad that that decision has resulted in a lot of people seeing it. I love you and I need to have some time to process this and think about how we might address it together."

(Being aware of your **window of tolerance** is important. It's okay to take a break so that you can return to the conversation when you aren't in your Downstairs Brain.)

Now, here's how we *proactively* reinforce or follow up on this:

Parent: Thank you for sharing with me that you feel embarrassed about your behavior choices. I'm still pretty upset, but believe it or not I was a kid once and I did some stupid things too. I made some choices that were not so good, only mine weren't on display for everyone to see! I don't want you to feel like you're alone in this.

(Remember, just because we are validating their feelings, it doesn't mean that we have to be okay with their behavior choices.)

"I wonder what led you to drink alcohol at that party?"

(Using wondering as a **verbal response** is a great way to help your teen consider what led them to make the choice that they did—it's developing their **abstract thinking** skills.)

Teen: Everybody else was doing it, so I thought I would too.

"I get it. We've all been there. Peer pressure is hard to deal with. Were any of your friends not drinking alcohol?"

"Yeah, a couple of them."

"I wonder what made it easier for them to not drink?"

"I don't know."

"I wonder if you felt like you needed to prove something to somebody."

"Maybe, but also I just wanted to try it."

"Thanks for being honest with me. It's hard to be a kid sometimes when there are so many things tempting you, especially when your brain isn't the best at making decisions yet."

"Yeah, believe me."

"I understand how upset you are that this photo of you was posted. Of course, you feel angry and embarrassed."

"So many people have seen it—they are all laughing at me."

"This is really hard, and I am sorry this happened."

"I will never be able to show my face again."

"I'm sure it feels like that. Unfortunately when things get posted online it can be hard to get them removed. It's important, when you are making choices about your behavior, to ask yourself if you would be happy with the whole world seeing it."

"Do we have to talk about this?"

"I know this is hard. It's not nice for us to have to sit and talk about choices we have made that we aren't really proud of."

(Sometimes **sitting in the yuck** is the best thing to do.)

"I wonder if you were invited to the party again would you do anything differently?"

(Here we are starting to help our teen **problem-solve** and **think abstractly**.)

"Well, I probably wouldn't even go."

"Yeah, that's a possibility. I wonder if there's a way that you could go and avoid the situation that happened this time?"

"Well, yeah, I could just not drink any alcohol and then I wouldn't have passed out and I wouldn't have had my picture all over a social media."

"Yeah, it seems like that's a pretty good answer. How do you feel about that?"

"Well, I feel stupid that I even did it in the first place."

"I hear you. Like I said earlier, we will do stupid things when we're growing up, especially at your age when your brain is still making

all these big changes; we can make some really bad choices. Why do you think we put these rules in place for you?"

"Yeah, I know."

(FYI, this might be our teen's way of saying "I'm sorry.")

"I think it would be helpful for us to go over our rules and expectations so that we can avoid this happening again. This is also a good time for us to talk about the risks when personal information about us gets posted online."

(Despite their age, our kids still need **structure**.)

In this particular situation, you might want to think very carefully about your choice of consequences, bearing in mind that your child is already receiving a pretty hefty **natural and logical consequence**. For some parents this may be enough. Others may feel that a more appropriate consequence is to ground the teenager and restrict access to friends' houses for a while.

If this is something that you decide to do as parents and caregivers, we encourage you to engage your teenager in other activities that take place within the family home during the time that they are grounded. This allows you to maintain your relationship with your child as well as help them feel connection during a time when they are already feeling sad.

You must choose what is right for you and your family. All we suggest is that you prioritize relationships in your decisions.

DIGITAL DISRUPTION
SCENARIO #17

"I used my 17-year-old son's tablet and noticed several pornography sites bookmarked. What should I do?"

DTAP® Concepts to Think About		
Downstairs Brain	**Midbrain**	**Upstairs Brain**
Window of Tolerance	Verbal Responses Sitting in the Yuck	Problem-Solving and Abstract Thinking Natural and Logical Consequences

In the moment, here are some things you need to remember before you respond:

Any parent who unexpectedly finds pornography on their child's device will feel a sense of shock. While your teen is becoming more independent and requires privacy, gently addressing what you have seen using **verbal responses** will benefit your child. Have a non-confrontational discussion with your son. This conversation will undoubtedly feel awkward, no matter how much you practice. Every parent fumbles over words when they are nervous, and this conversation is no exception. There will be a lot of **sitting in the yuck**.

Sexuality plays an integral role in the identity and development of teenagers, and it is expected that they will be curious about sex. This includes questions about masturbation, sexual fantasies, and sexual orientation. For some teens, they look to pornography for those answers. For others, mood management—managing emotional stress or discomfort, feeling lonely or impulsivity-related behaviors to seek sensation—might be why they access pornography. Viewing this type of online content can become habitual. While it is difficult to classify what is a habit or an addiction, the frequency of viewing porn, the content, and the amount to which it interferes with overall health, such as sleep, should be considered. This would be an opportunity to discuss **natural and logical consequences** with your teen.

Because teens are still developing **abstract thinking**, they will need you to emphasize that pornography is unrealistic, disrespectful, and manipulative. You may want to include information in your discussion with your child about viewing pornography, resulting in a more sexually permissive attitude, and with that comes risks such as sexually transmitted infections and unwanted pregnancy.

You might be tempted to take away all technology such as devices, apps, and internet access that could be a way to access pornographic material. This will not work. Your teen has grown up communicating through social media, and without some access, they can feel isolated. In addition, the goal is to help support your teen in using technology wisely while they have you to guide them—this is how we develop their own **problem-solving** skills.

Ultimately we want our teens to have digital media literacy and citizenship. *Digital media literacy* is the ability to critically, effectively, and responsibly access, use, understand, and engage with media of all kinds. *Digital citizenship* is the ability to navigate our digital environments in a way that's safe and responsible, and to actively and respectfully engage in these spaces.

Here's what to say to manage this, *in the moment*:

Parent: There is something I want to talk to you about, and please know that I am not mad at you. I noticed some pornography sites bookmarked on your tablet.

(If you are feeling mad, you are outside of your **window of tolerance** and potentially in your Downstairs Brain, so now is not the time to have this conversation.)

Teen: Oh my gosh, Mom, seriously, I do *not* watch porn!

"I do not want to debate that. Maybe you have looked at them and maybe you haven't. It is very normal to be curious about sex, and it is easy to access information online. What is most important is that you understand what pornography is and what it is not."

"Mom, I literally already know what porn is."

(You are going to have to **sit in the yuck**; it's normal that your teen does not want to have this conversation.)

"I suppose you know a lot, but a few things need clarification. Pornography is VERY different from what real people do in the privacy of their bedrooms."

"Do we have to talk about this right now? It is super awkward."

"Yes, we do. Yes, it is awkward for me as well. Hey, I am feeling nervous too. We can both feel nervous together. However, this is important, and if you have questions, I want you to know that you can ask me or someone who can give you facts. You need to understand that with pornography, sex is casual and purely physical, when in real life, sex is relational and emotional."

(While your teen may not want to hear it right now, it's important that you help them to **think abstractly** about the content they are viewing.)

"Okay, Mom, I get it."

"Thank you for hearing me out. I would not be doing my job if I did not talk to you about this. And just be careful that the time you spend online, whatever you are doing, isn't taking up time for real-world activities, or messing with your sleep."

(This is a gentle reminder about the **natural and logical consequences** of too much screen time.)

Now here's how we *proactively* reinforce this:

Parent: Hey! I know we had that really awkward chat last week about pornography.

Teen: Do we have to do this again, Mom?!

"I know, I promise I'm not going to make you talk about it. I just want you to know that I care about you and that you're a normal

teenager. And just remember, I'm here if you want to talk about anything. I won't judge you. I'm here to help if you need it."

"Okay, Mom. Thanks."

Sometimes it doesn't take much, just a gentle reminder to our kids that we see them, hear them, and care about them. Also, that we are paying attention to what they are doing online, not because we want to get them into trouble but because we want to keep them safe.

Our ability to have difficult, uncomfortable conversations will only serve to make it easier for them to come to us when they really need us.

Breathe through the awkwardness; it won't last long.

DIGITAL DISRUPTION
SCENARIO #18

"My 18-year-old daughter confessed that she got into some trouble on a gambling app, and she is over twenty-five hundred dollars in debt. I don't know what to do."

DTAP® Concepts to Think About		
Downstairs Brain	**Midbrain**	**Upstairs Brain**
Structure and Nurture	Verbal Responses Sitting in the Yuck	Problem-Solving and Abstract Thinking Natural and Logical Consequences

In the moment, here are some things you need to remember before you respond:

Our older teens are exposed to many things now that we probably weren't exposed to, or had easy access to when we were their age. Not only that but social media normalizes betting on sports, and much sports commentary is organized around gambling (how many points a player will score or how many yards a player will run in a game). Sports-betting platforms are like online games, with bonus points and rewards. This makes it familiar to teens who don't yet have the **abstract thinking** skills to really understand the difference between playing for fun and playing for money and that significant charges can be accrued quickly.

There are already substantial **natural** (a debt needs to be paid) and **logical** (parental controls added to technology and restrictions placed on bank accounts) **consequences** of your teen's actions. As a family you will need to decide how the debt is paid. (It's unlikely your teen has $2,500 readily available, and to avoid interest you may wish to pay it off. If that is the case, it is recommended that the teen be asked to pay back the money to you or to do tasks for you that add up to the value of $2,500.) This conversation will mean **sitting in the yuck** with our teen while also setting limits through **structure**. And don't forget, a little **nurture** always helps, no matter our age.

Gambling is highly addictive and almost impossible to resist. Why? Gambling is intermittent reinforcement, which means you never know when you are going to "hit," and that makes it hard to walk away. This stimulates the brain's reward center, and when the stimulation becomes excessive, it can trigger pathways to addiction. We can use our **verbal responses** to explore this with our teen in a way that is empathic but also maintains our expectations.

Once the initial issue has been resolved, we will need to encourage our teen to **problem-solve** with us to avoid a situation like this in the future.

Here's what to say to manage this, *in the moment*:

Parent: Am I hearing you clearly? You are in that much debt from gambling?

Teen: Yes, I am, and I am so sorry. I literally do not know how that even happened. I only bet $5 or $10 at a time.

"Okay, let me take a deep breath and think about this for a few minutes. Then, we can talk because I have many questions."

(You must remain calm while processing this with your older teen so you can remain empathetic and offer co-regulation if needed—if you need a break, or to tap out until you are calm, that's okay.)

"Okay, let's sit down and discuss this now. This is serious."

"I know it is. All of my friends use these apps."

"I understand how common it is, and since you love sports so much, I can see how this makes it even more fun, but now you are faced with a very big financial consequence."

(You are helping your teen begin to understand the **natural consequences** of their actions and promoting the **abstract thinking** skills needed to avoid this happening again.)

"The money added up fast."

"Yes, it did. I am sorry you are in this situation. This cannot continue to happen. I do not, and you certainly do not, have that kind of money to take risks with on betting. I wonder how we are going to figure out how to pay for this debt."

(We are providing empathy through our **verbal responses**, but we are not dropping the **logical consequence**.)

This would be a good time to talk about options for repaying the debt that work for your family.

"Now I would like to talk to you about your access to these apps and websites. I am not going to take away your phone or other devices, but I think it is warranted that I add some parental controls to your digital media, until I feel like you are making better decisions. Is that fair?"

(By asking for your teen's thoughts about your suggestion, we are actually balancing **structure and nurture—nurture** can look and sound different with an older teen.)

"Yes, it is fair; I actually want you to, so I am not tempted to use those apps."

"Well, I can help you with that accountability, but ultimately, you will have to develop skills to handle that independently. For now, I'd like to access your bank information so I can monitor how you are using your money."

(Reminding your teen that they will have to do things independently in the future will encourage them to engage in **problem-solving**.)

"Okay."

Now here's how we *proactively* reinforce this:

Parent: Hey, let's do a check-in regarding your internet use.

Teen: Oh Mom, do we have to?

"Yep. I know, it's not ideal but unfortunately that gambling situation got us in a big mess, didn't it?"

(Here you are **sitting in the yuck** with your teen.)

"Yeah, I know."

"So, I want you to know that I'm proud of you for managing to stay away from those sites and for doing what you need to do to pay off the debt."

"Thanks, Mom."

"I'd like to know if you've had a chance to think about how you're going to avoid this in the future?"

"Yeah, I've been thinking on it a bit. I think I am just going to avoid those websites and apps from now on."

"And what happens if your friends want to place some bets?"

"Ugh, yeah that might be hard. I guess I'm just going to not engage in that."

"Okay, I'm glad you are thinking about it. I'm always here. If you feel tempted, you can just send me a message and I will help remind you of what happened this time."

"That's a good idea."

We hope that this **natural and logical consequence** will help our teen avoid problematic choices in the future, however, it's important that we maintain an adequate level of support and supervision as their brain continues to mature and develop the skills needed to make smart decisions independently.

TOP TECH TIPS FOR THE LATE TEEN YEARS

1. While our older teens are becoming more independent, the content they can access online contains real risks that can have long-lasting consequences. It is still important that we set limits and expectations with them.
2. You will likely have to have many conversations regarding their digital media use, some of which will be uncomfortable for everyone, but if you don't talk to them about it, they will only learn from what they see and hear online.
3. Balancing independence with frequent check-ins is important.

QUICK SWAPS TO TRY

Instead of this...	Try this...
Ignoring concerning behavior because you don't want to invade privacy	Talk to your teen and teach them digital literacy, while also asking them to teach you about the things they like doing online.
Having one conversation about online activity	Have frequent, small check-ins to provide reminders.
Staying angry about online activities or behaviors	Create a safety net so that they know they can come to you if they experience something inappropriate or concerning online.

Small steps now help set our kids up for a lifetime of success. It's ok if mistakes happen but it's not ok for us not to address them, even if we feel uncomfortable doing so. Remember, our older kids still need a wiser mind from time to time.

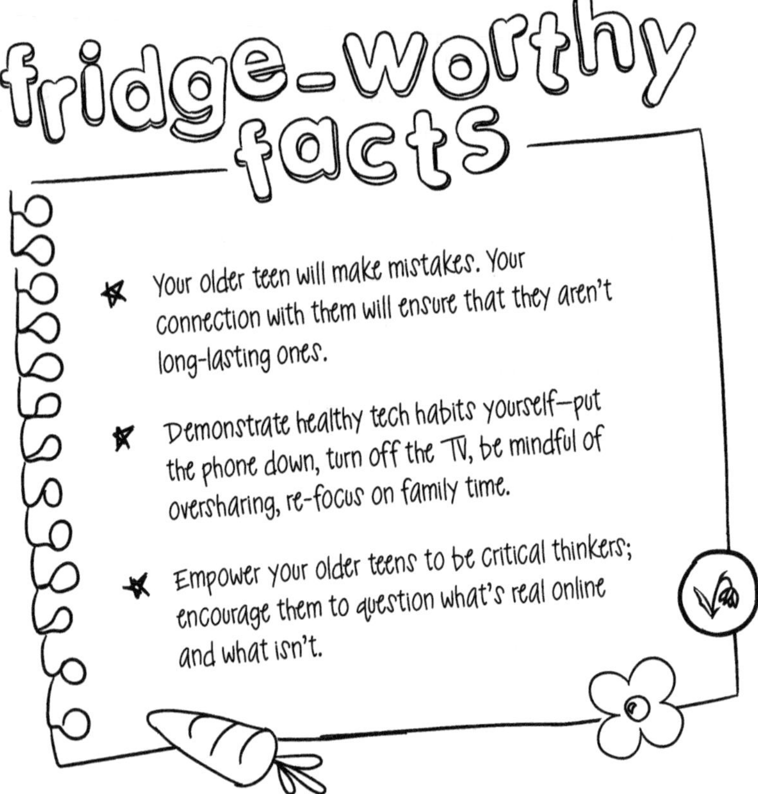

Part 3

THE FUTURE OF CHILDHOOD IN A DIGITAL WORLD

Chapter 10

GUIDING OUR CHILDREN TOWARD BALANCE, BELONGING, AND WELL-BEING

"Keeping Childhood at the Heart of Technology"

Childhood is changing—rapidly. Devices are everywhere, and digital fluency now begins before a child can tie their shoes. But while the tools may be new, a child's core needs have remained constant: connection, co-regulation, and yes, correction. As we look ahead, the question isn't how do we remove screens; it's how do we protect and foster healthy development while living in a digital world?

As we have pointed out throughout this book, technology isn't going away. In fact, when used thoughtfully, it can be a powerful tool for learning, creativity, and even connection. But it can't replace what brains are wired for: real relationships, embodied play, emotional attunement. These are the ingredients that build strong neural pathways and resilient kids.

If we want to raise children who can thrive in both physical and digital spaces, we have to be intentional. That means thinking beyond rules and restrictions, and instead focusing on setting realistic expectations and limits, while also maintaining relationships.

The approach we have given you within this book—the DTAP® approach—isn't just a list of tactics. It's a set of scientifically informed strategies that you can use as your child grows to ensure they become emotionally grounded, relationally connected, and capable of navigating life—online and offline—with awareness and integrity. By understanding and applying the DTAP® concepts to your digital-disruption challenges, you're no longer reacting out of fear. You're responding with insight. You're reading your child's cues, honoring the brain's developmental stages, and guiding—not micromanaging—the use of technology.

Over time, this builds:

- self-regulation instead of reliance on external distractions,
- empathy and communication skills instead of digital detachment,
- critical thinking rather than passive consumption, and
- confidence in navigating complex social and emotional worlds.

The long-term result? Children who don't just "cope" with the digital age but who know how to set boundaries, ask for help, and use technology as a tool—not a crutch.

Here's the hopeful part: You don't have to be perfect to make a difference. In fact, the most powerful changes often come from small, consistent actions done with intention.

You already have what it takes to lead your child toward a more balanced and connected life. It starts with:

- being curious, not controlling;
- modeling your own healthy digital habits;
- offering presence more than pressure; and
- seeing tech not just as a risk, but as a relationship opportunity.

By showing up with love, curiosity, consistency, and predictability, you become the anchor your child needs in a fast-moving world. You help them feel safe enough to explore and strong enough to return when they get overwhelmed.

A CLOSING REFLECTION

As we move into an increasingly digital future, let's not forget what childhood is meant to be: a season of connection, play, growth, and discovery. No app or platform can replace the power of a caregiver who sees, hears, and responds to a child with intention.

Remember, it's not about perfection; it's about presence. It's not about rejecting technology but building a foundation strong enough to thrive within it.

And that begins with us.

REFERENCES

George M. (2019). The Importance of Social Media Content for Teens' Risks for Self-harm. *The Journal of Adolescent Health: Official Publication of the Society for Adolescent Medicine*, *65*(1), 9–10. https://doi.org/10.1016/j.jadohealth.2019.04.022

Haidt, J. (2024). T*he Anxious Generation.* Penguin Press.

Hagen, M. (2024). What is Co-Regulation? Child Mind Institute. https://childmind.org/article/what-is-co-regulation

Hershler, A., Hughes, L., Nguyen, P., & Wall, S. (Eds.). (2021). *Looking at Trauma: A Tool Kit for Clinicians* (Vol. 23). Penn State University Press. https://doi.org/10.5325/j.ctv1wmz3qr

Hughes, D.A. (2006). *Building the bonds of attachment: Awakening love in deeply troubled children* (2nd ed). Jason Aronson.

Hughes, D. A., & Baylin, J. (2012). *Brain-based parenting: The neuroscience of caregiving for healthy attachment.* WW Norton & Company.

Isenberg, P & Quisenberry, N. (2002). Play Essential for All Children. A Position Paper of the Association for Childhood Education International.

Konrad, K., Firk, C., & Uhlhaas, P. J. (2013). *Brain development during adolescence: neuroscientific insights into this developmental period.* Deutsches Arzteblatt international, 110(25), 425–431. https://doi.org/10.3238/arztebl.2013.0425

Koplow, L. (2002). *Creating Schools That Heal: Real Life Solutions.* Teachers College Press.

National Scientific Council on the Developing Child (2007). *The Timing and Quality of Early Experiences Combine to Shape Brain Architecture: Working Paper No. 5.* Retrieved from www.developingchild.harvard.edu.

Suhag, K., & Rauniyar, S. (2024). Social Media Effects Regarding Eating Disorders and Body Image in Young Adolescents. *Cureus, 16*(4), e58674. https://doi.org/10.7759/cureus.58674

Tronick, E. (2009, February 25). *Still Face Experiment: Dr. Edward Tronick* [Video]. YouTube. https://link.edgepilot.com/s/1137553f/wJwLSnXvvUWck85OK-2gQA?u=https://www.youtube.com/watch?v=apzXGEbZht0

Vogels, E.A., Gelles-Watnick, R., & Massarat, N. (2022, August 10). *Teens, Social Media and Technology*. Pew Research Center. https://pewrsr.ch/3pc9pXn

PROFESSIONAL HELP AND RESOURCES

Over the next few pages, you will find our Top Ten Tech Tips for protecting your kids and teens online, as well as professional help and resources that we recommend to help you keep your child and teen safe as they use screens, the Internet and social media. You will find general support for ensuring the tech your family uses is set up to safeguard children, as well as links to websites and organizations designed to protect and respond to specific concerns such as self-harm, gambling and sexual exploitation.

Remember: Safety isn't just about blocking risks—it's about building skills, trust, and connection so kids can navigate the digital world with confidence and know that they have someone to turn to when things go wrong.

CHADDOCK'S TOP TEN TECH TIPS:

1. Connection Comes First

Practice what you have learned in this book by prioritizing connection. Kids who feel safe with you will tell you when something's wrong.

2. Use Technology Together

Sit with your child while they're online when possible or if you have an independent teen ask questions, explore together, and model how to think—not just what to avoid.

3. Set Clear, Consistent Boundaries

Define expectations for when devices can be used, where (keep use in shared spaces), and what apps or platforms are allowed. And make sure you follow through. Consistency builds security.

4. Delay Social Media Access

The longer you can delay high-risk platforms, the better. Remember, kids need time to develop the Upstairs Brain skills needed to make good decisions, think abstractly and utilize self-control.

5. Teach Digital Smarts

Help kids understand that not everything online is true, apps are designed to keep them hooked, and what they post can last forever.

6. Use Parental Controls (But Don't Rely on Them)

Controls can limit screen time, filter content, and monitor activity—but they work best alongside strong relationships and communication.

7. Keep Devices Out of Bedrooms at Night

Late-night, unsupervised use increases risk and disrupts sleep. Create a central charging spot for all devices.

8. Teach "Pause Before You Post"

Encourage kids to ask: Is this kind? Is this safe? Would I want others to see this? Help them think about the potential consequences of their actions.

9. Practice What to Do in Tough Situations

Give kids clear language: "I'm not comfortable with that." "I'm going to tell a trusted adult." Remind them they won't get in trouble for asking for help.

10. Model Healthy Tech Use

Your habits matter. Have device-free conversations, set boundaries and limits with your own screen time, and presence and attention. You are their most powerful example.

GENERAL GUIDANCE, ADVICE AND TECH SUPPORT

Bark Website

The Bark website does more than just provide apps to help monitor your kids online activity (although they are a fantastic tool for this purpose). You will find guidance on Internet safety, a full list of commonly used online slang terms, including emojis, acronyms and abbreviations (we highly recommend you become literate in the language your kids and teens are using!)
www.bark.us

Common Sense Media Website

Common Sense Media website has an enormous amount of guides for parents, caregivers and educators on everything from Movies to Social Media to Video Games. It also includes information on technology use through the ages from pre-schoolers to teens.
www.commonsensemedia.org

Internet Safety 101 Website

From example digital/tech family contracts to statistics to guides on A.I. and Cybersecurity – the Internet Safety 101 website is a great resources for families wanting to know and understand more.
www.internetsafety101.org

SELF-HARM AND SUICIDAL BEHAVIORS

If you are concerned that your child or teen is thinking about or engaging in self-harm behaviors or having suicidal thoughts, first and foremost, try to remain calm. We understand that this can be a scary time for both adults and children, but our kids need us to be the safe, regulated person they can turn to. It's ok to express your worries but it's important that you do so in a way that won't increase the fear your child is likely already experiencing.

Talk calmly and explain to your child that you want to make sure they are safe.

Encourage collaboration, ask for their input and seek professional help.

Below you will find several resources, accessible by phone and online, with professionals available to support you and your child.

988 Suicide and Crisis Lifeline:

Self-Harm Support
Dial 988
Languages: English, Spanish
Hours: 24/7

Crisis Text Line:

Text CONNECT to 741741 for free, confidential support from a trained volunteer Crisis Counselor, available 24/7.
www.crisistextline.org/topics/self-harm/

Young Minds Website

Resources and Guides on Self-Harm Behaviors
www.youngminds.org.uk/parent/
parents-a-z-mental-health-guide/self-harm

Mental Health America Website

Information and Resources
https://mhanational.org/resources/self-injury-and-youth

SEXUAL EXPLOITATION

If you're worried your child or teen might be impacted by online sexual exploitation and abuse, here are some things to look out for:

- Sudden change in mood or emotional state.
- Evasiveness about online activity, such as minimizing screens, hiding accounts and apps
- Withdrawal from real-life friends
- Hyperfocus on the online environment
- Receiving random gifts from strangers
- Use of language not used before

When your child or teen is using a public platform, it is recommended that privacy settings be used and adjusted to control who can send direct messages, make friend requests, and access personal information.

Look for parental control features that monitor your child's activity, including servers they join, people they interact with, and recent messages. You can also enable Two-Factor Authentication (2FA) for extra security. Educate your child or teen about blocking unknown users and what to do if someone shares inappropriate content.

We also recommend you talk to your child or teen about the following:

- Do not use your real name or any kind of suggestive screen name – this lowers the risk of unwanted attention.
- Be wary of people paying you attention or giving you compliments online, especially if you don't know who they are.
- Don't share personal information (including name, age and location) and be wary of people who ask personal questions.
- Do not share nude or intimate pictures.
- Remember that people aren't always who they say they are.
- Never arrange to meet someone you met online in-person.

- Always talk to a trusted adult if you run into issues online, even if it makes you uncomfortable. Adults can help keep you safe.
- Save any conversations or pictures sent to you as they might be used for evidence.

How to report online sexual exploitation and abuse:

Know2Protect Tipline:

Dial 1.833.591.KNOW(5669)
report.cybertip.org

GAMBLING

If you suspect your child or teen might be involved in gambling, use the resources below to help you get the help and support you need.

GambleAware

Dial 0808.8020.133
Free advice, tools and support to keep people safe from gambling harm.
www.gambleaware.org

National Council on Problem Gambling

The National Council on Problem Gambling provides a range of resources, including answers to commonly asked questions,

a gambling behavior self-assessment and information about treatment.
Dial 1.800.MY.RESET
www.ncpgambling.org

EATING DISORDERS OR PROBLEMATIC EATING BEHAVIORS

National Eating Disorders Association

Information and resources to help you support your child or teen who may be experiencing disordered eating behaviors.
www.nationaleatingdisorders.org

ABOUT CHADDOCK

Chaddock is a nonprofit organization based in Quincy, Illinois, with a nearly 200-year history of serving children and families. It is internationally recognized for its expertise in helping children who have experienced significant trauma, abuse, neglect, and attachment disruptions.

Chaddock's mission is to strengthen children and families through innovative, trauma-informed and attachment-based services. The organization provides a comprehensive continuum of care for children from birth through age 21, including residential treatment, a special education school, community-based programs, and in-home services.

Grounded in research and real-world practice, Chaddock's approach emphasizes relationships as the foundation for healing. Their work integrates developmental trauma and attachment science to support emotional, behavioral, and relational growth in children while equipping caregivers and professionals with practical tools and training.

In addition to direct services, Chaddock extends its impact through professional development, consultation, and educational resources—sharing its expertise with educators, clinicians, and organizations across the country and around the world.

To learn more about Chaddock, the Developmental Trauma and Attachment Program (DTAP®), training opportunities and resources for families and professionals, visit www.chaddock.org

ABOUT THE AUTHOR

Kirsty Nolan has a singular mission in this world: To help parents and caregivers be their best, because she knows firsthand how difficult that job can be and how much is riding on getting it as right as humanly possible.

Kirsty was raised in the heart of Suffolk, in the U.K., and in her own words had a "very good childhood." She would complete her undergraduate degree from the University of Wales and her master's degree from the University of Leicester.

From her earliest days, Kirsty has always been compassionate about helping those who are vulnerable. She innately knew that loving one's neighbor as oneself was the right thing to do and as best as she has been able, she has lived her life by that credo.

Thankfully she has found an organization that not only lives by that golden rule but encourages its people to live by it as well. She has been employed at Chaddock since 2014. Chaddock is an organization that has dedicated itself to helping families and children for almost 200 years.

Rebooting Childhood: How to Balance Screen Time, Build Relationships and Help Kids Thrive is Kirsty's first book.

She lives in O'Fallon, Illinois with her husband, Chris, her 2-year-old daughter and 16-year-old stepdaughter.

www.ingramcontent.com/pod-product-compliance
Lightning Source LLC
LaVergne TN
LVHW020659110826
845149LV00012B/2052